Secret Weapons

Secret Weapons

Defenses of Insects,
Spiders, Scorpions, and
Other Many-Legged Creatures

THOMAS EISNER

MARIA EISNER

MELODY SIEGLER

The Belknap Press of
Harvard University Press

Cambridge, Massachusetts
London, England
2005

Library of Congress Cataloging-in-Publication Data
Eisner, Thomas, 1929–
    Secret weapons : defenses of insects, spiders, scorpions, and other
many-legged creatures / Thomas Eisner, Maria Eisner, Melody Siegler.
      p. cm.
    Includes bibliographical references and index.
    ISBN 0-674-01882-6
    1. Insects—Defenses.   2. Arachnida—Defenses.   I. Eisner, Maria.
II. Siegler, Melody.   III. Title.
QL496.E42 2005
595.7147—dc22   2005041042

Contents

Order COLEOPTERA

No pursuit at Cambridge . . . gave me so much pleasure as collecting beetles . . . I will give you proof of my zeal: one day, on tearing off some bark, I saw two rare beetles, and seized one in each hand; then I saw a third and new kind, which I could not bear to lose, so I popped the one which I held in my right hand into my mouth. Alas! It ejected some intensely acrid fluid, which burnt my tongue so I was forced to spit the beetle out, which was lost, as was the third one.

Charles Darwin, *Life and Letters,* 1874

Prologue

This book is intended as an illustrated guide to the defenses of insects and their kin, to the strategies that enabled present-day arthropods—insects and their relatives such as spiders, scorpions, and centipedes—to achieve preeminence on land. The book contains 69 numbered chapters, short narratives focused on individual defensive strategies, as exemplified by selected arthropods or arthropod groups. We did not strive for comprehensiveness in our treatment of the subject. We chose stories that could be told in pictures, which meant being guided to some extent by what was available to us in our photographic files. It also meant restricting ourselves to a North American cast of characters, given that it was with North American arthropods that we did most of our research. Some examples are given of defensive scenarios that play themselves out in Australia, Europe, and Asia, but these are relatively few. The fact is, however, that one needn't draw examples from the entire world to prove that defense is at the root of the evolutionary success of arthropods. Examples from a single geographic area suffice to make the point.

It was of central concern to us to provide, through our choice of examples, some feeling for the diversity of arthropod defenses, and in particular for the range of chemical defenses, since these are the principal ones found in arthropods. We therefore chose to provide coverage of glandular defenses, of examples involving the use of sprays, oozes, sticky coatings, or froths, and of nonglandular defenses, dependent on the use of enteric fluids, feces, or systemic toxins. We chose also to deal with the question of how arthropods come to possess their defensive chemicals, whether by endogenous synthesis or by appropriation from an exogenous source, and with how protection may be provided by visual means, as through camouflage, mimicry, or warning coloration. Further, we decided to consider venoms, especially those administered by stinging, used for protective purposes in social contexts by the likes of ants and honey bees. And finally, we thought it would be appropriate to give examples of how insects protect their offspring, sometimes through biparental action, and of how defense may be at issue in the communicative "dialogue" between males and females during courtship.

Our book is a first of its kind. A photographic introduction to the defensive capabilities of arthropods has not previously been produced. The book is intended for a broad audience, for naturalists of all kinds, for the teacher as well as the student, the hands-on researcher as well as the observer, and ultimately for all those to whom nature has never ceased to be a source of wonder.

The book is organized in such a way that it can be read piecemeal. The individual case studies are self-contained, so the book can be used as a reference guide. We chose examples from most orders of insects, as well as from all major classes of noninsectan terrestrial arthropods—the whipscorpions, true scorpions, centipedes, millipedes, and spiders—and the sequence of presentation is taxonomic. Noninsectan examples are presented first, followed by the insects, grouped by order.

For each arthropod presented, we give both the Latin binomial name, and the common name, when one exists (we use the common names officially sanctioned by the Entomological Society of America or those mentioned in field guides and other authoritative treatises). When common names are available, these are given, preceded by the article "the"; for instance, the Mexican bean beetle. When such names are lacking, species are referred to according to their taxonomic position; for example, a scarabaeid beetle.

Throughout the text, we provide chemical formulas for the substances that play mediating roles in the various defensive strategies discussed. These chemicals are numbered consecutively in the order in which they appear in the text, and are referred to by these numbers when they are mentioned subsequently in the narratives.

Our aim, in the preparation of the bibliographies at the ends of chapters, was to provide the reader with entry points to the existing literature, rather than with a comprehensive coverage of all that has been written on the various topics presented. Our dependence on available photos led us to give preferential coverage to some of our own investigations and to cite primarily our own publications. There are obviously many other studies that could have been mentioned, and that we would have highlighted had our treatment been all-inclusive, and had the photographic documentation been available to us. Readers interested in looking beyond these pages are urged to access the publications of other investigators, including among many others the following, who as leading chemical ecologists have contributed directly or indirectly to the study of chemically based defensive interactions of arthropods: J. R. Aldrich, A. B. Attygalle, M. Ayasse, I. T. Baldwin, J. X. Becerra, M. R. Berenbaum, G. Bergström, M. S. Blum, M. Deane Bowers, W. S. Bowers, L. B. Brattsten, R. Brossut, L. E. Brower, K. S. Brown, W. E. Conner, J. C. Daloze, J. W. Daly, K. Dettner, M. Dicke, D. E. Dussourd, D. L. Evans, H. M. Fales, P. Feeny, W. Fenical, W. Francke, T. Guilford, M. T. Hartman, C. D. Harvell, M. E. Hay, J. G. Hildebrand, M. Hilker, B. Hölldobler, T. M. Jones, J. Kubanek, W. S. Leal, U. Maschwitz, J. Meinwald, B. P. Moore, D. E. Morgan, K. Mori, A. Nahrstedt, R. Nishida, C. H. Noirot, J. M. Pasteels, M. Pavan, J. Pickett, G. D. Prestwich, W. Roelofs, M. Rothschild, C. A. Ryan, H. Schildknecht, J. O. Schmidt, D. Schneider, F. C. Schroeder, S. Schulz, R. M. Silverstein, K. Slàma, S. R. Smedley, D. W. Tallamy, J. S. Trigo, W. R. Tschinkel, J. H. Tumlinson, R. T. Vander Meer, D. W. Whitman, and E. O. Wilson.

As we stress throughout the text, many questions remain open in the field of arthropod defenses. The area is ripe with opportunity along a broad span of disciplines, ranging from the behavioral, ecological, and evolutionary to the neurobiological, molecular, and medicinal, and should therefore have the capacity to be as enticing to the naturalist as to the genomicist.

1

Class ARACHNIDA
Order UROPYGI
Family Theliphonidae
Mastigoproctus giganteus
The vinegaroon

Mastigoproctus giganteus.

The vinegaroon is aptly named. Its defensive spray, ejected when the animal is physically disturbed, contains acetic acid (**1**), the sour substance that gives vinegar its flavor. But whereas vinegar contains no more than a few percent acetic acid, the vinegaroon's spray consists of 84% acetic acid, doubtless the highest concentration at which this compound is found in nature. The other components are, principally, a second acid, caprylic acid (**2**), and water.

1. Acetic acid   2. Caprylic acid

The spray is produced by two glands situated posteriorly in the abdomen, or the opisthosoma, as the abdomen is called in arachnids. The glands open close together at the tip of the small knob that forms the base of the flagellum, or whip, that projects from the rear of the animal. The knob functions as a revolvable gun emplacement. By rotating the knob and adjusting the posture of the opisthosoma, the animal can direct its spray in virtually any direction. The precision of its aim can be demonstrated by causing the animal to discharge on filter paper impregnated with an indicator dye such as phenolphthalein, which turns from red to white in response to acids. Grasp a vinegaroon by an appendage, and it will spray precisely in the direction of that appendage. The glands are large enough to hold fluid for a number of discharges.

Caprylic acid contributes to the effectiveness of the defense by promoting the spread and penetration of the spray at its target. Without caprylic acid, the acetic acid would not be nearly as quick in effecting the irritancy that is so typical of the spray. The vinegaroon is a nocturnal predator. So far as is known, it uses its spray strictly for defense and never for incapacitating prey. Ants and mice are among the several predators that have been shown to be repelled by the vinegaroon's spray.

The Uropygi make up an ancient group, consisting of only some 85 species worldwide, all placed within the single family Theliphonidae. The defensive glands are characteristic of the group and may well have contributed to the long-term survival of these animals. The secretion has been studied in only a few species, but appears to be acid-based throughout the group. The genus *Mastigoproctus* itself contains 12 species, from North America, the West Indies, and South America.

A vinegaroon discharging its acid spray toward a leg that is being pinched with forceps. The spray pattern shows up in white on paper impregnated with phenolphthalein solution.

REFERENCES

Eisner, T., J. Meinwald, A. Monro, and R. Ghent. 1961. Defense mechanisms of arthropods. I. The composition and function of the spray of the whipscorpion, *Mastigoproctus giganteus* (Lucas) (Arachnida, Pedipalpida). *Journal of Insect Physiology* 6:272–298.

Schmid, J. O., F. R. Dani, G. R. Jones, and D. E. Morgan. 2000. Chemistry, ontogeny, and role of pygidial gland secretion of the vinegaroon *Mastigoproctus giganteus* (Arachnida: Uropygi). *Journal of Insect Physiology* 46:443–450.

2

Class ARACHNIDA

Order OPILIONES

Family Cosmetidae

Vonones sayi

A harvestman

Vonones sayi.

Spider-like and encased in a hard carapace, *Vonones sayi* looks as though it might be venomous. But it is not. It has neither venom glands nor the means to inject venom. Instead, when disturbed, it mixes oral effluent with the noxious products of two exocrine glands, then brushes the mixture onto the enemy with its forelegs. Its chemical defense is highly unusual, shared only with closely related Opiliones.

3. 2,3,5-Trimethyl-1,4-benzoquinone   **4.** 2,3-Dimethyl-1,4-benzoquinone

The two defensive glands of *V. sayi* are situated at the antero-lateral margins of the prosoma (the part of the arachnid body combining the head and thorax). The glands are small, but they contain the defensive chemicals in concentrated form, as a paste. These chemicals have been analyzed and found to be two quinones: 2,3,5-trimethyl-1,4-benzoquinone (**3**) and 2,3-dimethyl-1,4-benzo-quinone (**4**).

The oral fluid used by *V. sayi* as a diluent of the secretion is free of solid matter. Like most arachnids, *V. sayi* does not swallow solids. It grinds up its food while holding it in the mouthparts, drenching it in regurgitated enzymatic fluid and imbibing the resulting liquid. *V. sayi* mobilizes this liquid when attacked. If, for instance, you hold the animal in forceps, it disgorges a droplet of clear liquid from the mouth. This droplet quickly disappears as the fluid is conveyed by way of two narrow clefts to the sites at the edge of the carapace where the glands open. There, the fluid forms two droplets, one over each gland opening. Almost immediately the two droplets begin to swirl and take on color, as the glands void some of their yellowish quinonoid paste. Thus armed, *V. sayi* is set for action. If you persist in your hold, the animal dips its forelegs into the droplets, wets the tips, and brushes these against the forceps. The tips of the forelegs are hairy, like miniature brushes. The legs themselves are long and multi-jointed, and can be directed by the animal toward virtually any part of the body.

V. sayi's principal enemies are ants. In laboratory experiments ants have been shown to be highly sensitive to the repellent properties of benzoquinones and to be effectively "brushed off" by *V. sayi*. Spiders and other arthropods are likely to be similarly repelled. It has been

Top: *Vonones sayi,* with droplets of freshly formed benzoquinone solution clinging to its body. Bottom: *V. sayi* has picked up the solution with the tip of a foreleg, in preparation for brushing it against the attacker.

estimated that with replete glands and a full load of gut fluid, a *V. sayi* can fend off as many as 55 individual ant assaults.

Within the Opiliones, *V. sayi* belongs to the suborder Laniatores, together with some 2,000 other species. Benzoquinone production appears to be widespread among the group, and so may be the habit of using gut fluid for dilution of the secretion. Some Laniatores are known to produce phenols in their defensive glands, compounds that are also potently repellent. The common name "harvestman" applies not just to *V. sayi,* but to opilionids generally.

REFERENCES

Acosta, L. E., T. I. Poretti, and P. E. Mascarelli. 1993. The defensive secretion of *Pachyloidellus goliath* (Opiliones, Laniatores, Gonyleptidae). *Bonner zoologische Beiträge* 44:19–31.

Eisner, T., D. Alsop, and J. Meinwald. 1978. Secretions of opilionids, whip scorpions, and pseudoscorpions. In S. Bettini, ed., *Arthropod Venoms.* Berlin: Springer-Verlag.

Eisner, T., A. F. Kluge, J. E. Carrel, and J. Meinwald. 1971. Defense of phalangid: liquid repellent administered by leg dabbing. *Science* 173:650–652.

Eisner, T., C. Rossini, A. González, and M. Eisner. 2004. Chemical defense of an opilionid (*Acanthopachylus aculeatus*). *Journal of Experimental Biology* 207:1313–1321.

3

Class ARACHNIDA

Order OPILIONES

Family Sclerosomatidae

Leiobunum nigripalpi

A daddylonglegs

Leiobunum nigripalpi.

The opilionids of the genus *Leiobunum,* together with other species of the family Sclerosomatidae, while known as harvestmen like
opilionids generally, are also called daddylonglegs in North America,
in recognition of their long and slender appendages. Daddylonglegs
are familiar to most everyone with a naturalist bent. Children make
it a habit to pick them up. If a child grasps one by a leg, she is likely
to find herself holding that leg only, which the daddylonglegs de-

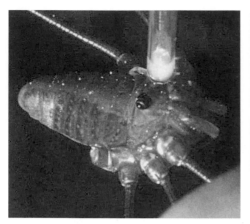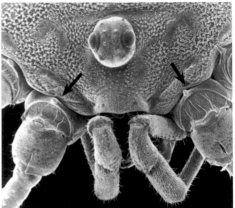

Left: An unidentified daddylonglegs being milked of its secretion. A glass capillary tube has been pressed against one of the gland openings, and the animal has discharged its secretion into the tube. Right: Top view of the front of the carapace of a daddylonglegs. The arrows point to the gland openings.

taches at the base to gain its freedom. She is also likely to notice that the autotomized leg continues to twitch for a while in her hand, giving the false impression that it is alive, and thus diverting attention from the fleeing opilionid.

Leg autotomy is the first line of defense for daddylonglegs. Individuals that you find in nature frequently are missing a leg or two, but are remarkably agile even when thus handicapped. Leg loss may occur also when the animal extricates itself from the exoskeleton during molting, but probably is chiefly due to encounters with predators.

Daddylonglegs have a second line of defense, which they use when they are grasped by the body. They have a pair of glands, opening dorsally near the anterior margin of the carapace, from which they discharge a potently odorous fluid when attacked. The fluid, which oozes over their body, is strongly repellent to such enemies as ants. The fluid varies in composition from species to species. In *Leiobunum nigripalpi,* a common North American daddylonglegs, the chief components of the secretion are *(E)*-4-methyl-4-hexen-3-one (**5**), 4-methylhexan-3-one (**6**), 4-methylhexan-3-ol (**7**),

5. *(E)*-4-Methyl-4-hexen-3-one   6. 4-Methylhexan-3-one   7. 4-Methylhexan-3-ol

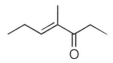

8. *(E)*-4-Methyl-4-hepten-3-one   9. *(E,E)*-2,4-Dimethylhexa-2,4-dienal

10. 1,4-Naphthoquinone   11. 6-Methyl-1,4-naphthoquinone

(E)-4-methyl-4-hepten-3-one (**8**), and *(E,E)*-2,4-dimethylhexa-2,4-dienal (**9**). Other species of *Leiobunum* produce related branched-chain alcohols, ketones, and aldehydes. Another daddylonglegs, but of a different genus, *Phalangium opilio,* produces two naphthoquinones (**10, 11**).

The long legs are themselves protective in daddylonglegs. They enable the animals to flee rapidly over vegetation, taking giant steps as they span the gaps between leaves and branches. The legs can also serve as defensive stilts. When disturbed, daddylonglegs sometimes stand rigid, with the body raised above the substrate. Ants on the prowl are likely to attack if they come upon the body of a daddylonglegs, but may not respond if they contact the legs alone.

The family Sclerosomatidae belongs to the opilionid suborder Palpatores, which is estimated to contain some 2,000–3,000 species, many still undescribed.

REFERENCES

Blum, M. S., and A. L. Edgar. 1971. 4-Methyl-3-heptanone: identification and role in opilionid exocrine secretions. *Insect Biochemistry* 1:181–188.

Eisner, T., D. Alsop, and J. Meinwald. 1978. Secretions of opilionids, whip scorpions, and pseudoscorpions. In S. Bettini ed., *Arthropod Venoms.* Berlin: Springer-Verlag.

Wiemer, D. F., K. Hicks, J. Meinwald, and T. Eisner. 1978. Naphthoquinones in defensive secretion of an opilionid. *Experientia* 34:969–970.

4

Class ARACHNIDA

Order SCORPIONES

Family Vejovidae

Vejovis spinigerus

The striped tail scorpion

Vejovis spinigerus.

Beware of animals that use offensive weapons for protective purposes. Such weapons are intended to kill, and may therefore be lethal also when used in defense. Humans are fearful of snakes, spiders, and scorpions, fully aware that these animals include species that are highly poisonous.

Scorpions are the oldest of the extant arachnids. They include some 1,200 species worldwide, of tropical and temperate distribution, and may occur in virtually any habitat from forests to deserts.

They are nocturnal hunters that feed on other arthropods. Most live at ground level, although some are arboreal. They seek shelter in the daytime, and may find their way into dwellings. Residents of areas where scorpions abound often take the precaution of shaking out their shoes in the morning before putting them on.

Scorpions characteristically have a pair of pincers out front and a long segmented tail at the rear. The stinging apparatus is at the tip of the tail, and consists of a swollen, capsule-like ampulla, containing the two glands, and a projecting, needle-sharp stinger, bearing separate gland openings near the tip. A scorpion uses its stinging apparatus with great efficiency. When hungry, it reacts quickly to prey in its vicinity, grasping the intended victim with the pincers while simultaneously thrusting the tail to inflict the sting. When using its stinger in defense, a scorpion similarly lashes its tail. Merely touching a scorpion can elicit a sting.

Scorpions have poor vision. They have a pair of eyes in the center of the head, and two or more along the margin on each side, but they make little use of these when hunting. Rather, they depend largely on touch, using the long hairs on their pincers to detect prey. When roaming, scorpions feel their way by use of a pair of unique comblike structures that project ventrally from the body, the pectines, with which they are able to gauge both mechanical and chemical stimuli.

All scorpions are venomous, although the potency of their venom differs greatly. *Vejovis spinigerus,* the striped tail scorpion, is one of the most common American species, particularly abundant in California and Arizona. As with the vast majority of scorpions, its sting is painful but short-lived in effect, likened to a bee or wasp sting, except in people allergic to the venom, who should promptly seek treatment if stung.

Only about 25 species of scorpions worldwide, all belonging to the family Buthidae, have venom potentially lethal to humans. In the United States there is only one life-threatening scorpion, the bark scorpion, or deadly sculptured scorpion, *Centruroides exilicauda* (formerly *Centruroides sculpturatus*). The species ranges from western New Mexico and Arizona to eastern California and southward into northern Mexico. It is unusual in being a climbing rather

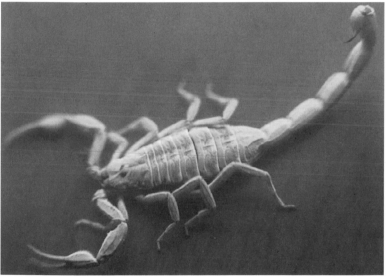

Top: An unidentified female scorpion (from southern Arizona) with young on her back. Bottom: A scorpion fluorescing in ultraviolet light.

than a burrowing species, and it makes its way occasionally into homes. Its sting can induce a multiplicity of nasty symptoms. These include severe pain and swelling where the stinger has penetrated, numbness even in areas at some distance from the sting, frothing at the mouth, convulsions, and respiratory paralysis. In Arizona, as late as the 1940s, this species was responsible for more deaths than all other venomous animals combined, including rattlesnakes, gila monsters, spiders, bees, and wasps. Nowadays, thanks to improved medical treatment, mortality has been drastically reduced. An antivenin exists, but medical opinion differs as to its effectiveness. The best strategy with scorpions is to exercise extreme caution.

Much has been learned over the years about the chemistry of scorpion venoms and their mode of action. Research has concentrated on venoms from medically important species. Venom contains water, salts, small molecules, peptides, and proteins. The effectiveness of the venom is almost wholly due to its large complement of diverse peptide toxins. Scorpion toxins, along with the so-called conotoxins from several species of the marine hunting snails of the genus *Conus,* have been a mainstay of basic and applied research into how membrane channels regulate ion concentrations inside individual cells. The toxins are stable, and effective when separated chemically, unlike many of the toxins that constitute spider or centipede venoms. The toxins bind "ion channels," the tightly folded proteins that span cell membranes and gate the movement of ions between the inside and outside of the cell. Binding occurs with great specificity, as if each particular peptide toxin was designed to compromise a particular type of channel. Over 200 distinct peptide toxins have been isolated and sequenced from 30 different species of scorpions. In *Centruroides* species alone more than 45 separate toxins have been identified. Most of these target neurons, but some affect other tissue types, in keeping with the widespread effect of certain species' venoms. The peptide toxins exert their effects either by blocking the passage of ions through the membrane channel or by interfering with the gating mechanism that controls the opening and closing of the channel. Of the many toxins characterized, some act specifically on vertebrates, others on insects, and still others on vertebrates, insects, and crustaceans, all three types sometimes oc-

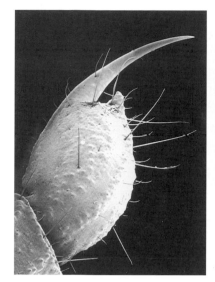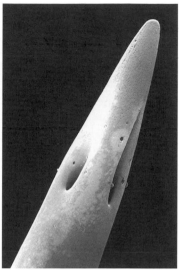

Left: The ampulla, with its stinger, of an unidentified scorpion. Right: The tip of the scorpion stinger, showing the separate slit-like openings of the two glands.

curring in the same venom. From a biological point of view it is of interest that scorpions appear to be insensitive to their own venom. They have in some instances been shown to tolerate venom or venom components at 100 to 1,000 times the concentration at which these materials affect the ion channels of other animals.

Venom production is inevitably costly to a scorpion because of the high protein content of the fluid, and it also takes time, with the result that scorpions risk becoming temporarily vulnerable following use of the stinger. It makes sense therefore that scorpions should economize on the use of venom. Indeed, one large species of scorpion was shown hardly ever to resort to venom injection as an adult, but to use its stinger routinely as an immature, when it is small relative to the size of its victims. Other scorpions, it seems, quite generally restrict venom injection to instances when they cannot subdue prey with the pincers alone.

An interesting recent finding is that when a scorpion first stings it injects a fluid that is chemically different from that which it injects with subsequent stings. The fluid injected first has been named "prevenom," and it may be intended specifically to induce localized pain in vertebrates such as mice, which commonly prey on scorpions. Prevenom is extraordinarily rich in potassium ions (K+), and it contains peptide toxins that block certain mammalian K+ channels. These toxins, in conjunction with the K+, appear to ensure that sensory neurons at the site stung continue to be active long after the sting, which explains why the site of a scorpion sting hurts for a while. The venom itself, when injected into mice, fails to elicit this protracted localized response.

The prevenom, by virtue of its high K+ content, can cause flaccid paralysis in insects. This alone could allow the scorpion to prevail over insectan prey. The venom, however, by virtue of its anti-insectan toxins can act as a powerful supplement to the prevenom.

How the scorpion manages the trick of segregating the prevenom from the venom in the sting apparatus, so that the two fluids can be injected separately, remains a mystery.

For naturalists intent on collecting live scorpions for study, the task is greatly facilitated by the odd fact that scorpions fluoresce under ultraviolet light. All that is necessary to spot scorpions in a desert at night is a portable "blacklight," which emits just the right ultraviolet wavelengths to make scorpions glow blue in the dark. It is impressive to discover how many scorpions are on the prowl at night, and how dangerous it must be for arthropods to risk coexistence with such hunters.

Young scorpions remain on the mother's back until they undergo their first molt. After shedding the exoskeleton, they become independent, and may live for several years.

REFERENCES

Inceoglu, B., J. Lango, J. Jing, L. Chen, F. Doymaz, I. N. Pessah, and B. D. Hammock. 2003. One scorpion, two venoms: prevenom of *Parabuthus transvaalicus* acts as an alternative type of venom with distinct mechanism of action. *Proceedings of the National Academy of Sciences USA* 100:922–927.

Legros, C., M. F. Martin-Eauclaire, and D. Cattaert. 1998. The myth of scorpion suicide: are scorpions insensitive to their own venom? *Journal of Experimental Biology* 201:2625–2636.

Levi, H. W., and R. W. Levi. 1968. *A Guide to the Spiders and Their Kin.* New York: Golden Press.

Olivera, B. M. 1997. Conus venom peptides, receptor and ion channel targets, and drug design: 50 million years of neuropharmacology. *Molecular Biology of the Cell* 8:2101–2109.

Possani, L. D., E. Merino, M. Corona, F. Bolivar, and B. Becerril. 2000. Peptides and genes coding for scorpion toxins that affect ion channels. *Biochimie* 82:861–868.

Rein, J. O. 1993. Sting use in two species of *Parabuthus* scorpions (Buthidae). *Journal of Arachnology* 21:60–63.

Russell, F. E., C. B. Alender, and F. W. Buess. 1968. Venom of the scorpion *Vejovis spinigerus. Science* 159:90–91.

5

Class ARACHNIDA
Order ARANEIDA
Family Oxyopidae
Peucetia viridans
The green lynx spider

Peucetia viridans, frontal view.

Spiders are easily recognized as such, but are rarely given the attention they deserve as important members of the natural world. They tend to inspire fear in humans, the proverbial arachnophobia, an attitude that has fostered ignorance worldwide about what is truly a most fascinating and useful group of animals. Spiders have survived the evolutionary onslaught of insects, and they have done so by becoming specialists that feed on insects. Spiders lie in wait for insects in the night, hunt for insects by day, and trap insects in their webs. Larger spiders also prey upon small vertebrates, including fish, lizards, birds, and mice. Although only some 35,000 species of spiders have been described (as compared to nearly a million species of insects), spiders are a diverse lot, as evidenced by their classification into over 80 families. To a budding naturalist in search of a lifetime commitment, the study of spiders can be heartily recommended. With little more than a stopwatch, a hand lens, and a good dose of curiosity, anyone can become an arachnophile. Work with spiders is bound to lead to discoveries.

Spiders typically kill by injecting venom. Their venom glands are associated with their fangs, the so-called chelicers, sharply pointed devices fashioned to perforate the insectan exoskeleton. Spiders waste no time in using their fangs, and if the prey is small, they may simply crush it or chew it, drenching it in digestive juices and imbibing the liquid food components. Larger prey are first subdued by the injection of venom. Wolf spiders, which stalk their prey, and jumping spiders, which pounce on their victims, inject their venom outright, the moment they make their catch. Orb-weaving spiders may be more circumspect. They may first envelop the prey in silk, and only then inflict their bites. By first wrapping the prey, they avoid being stung by the likes of bees and wasps, or being hit full blast by insects that eject defensive sprays.

Spiders, of course, can also inject their venom defensively. Given that their venom is intended to kill prey, it is not surprising that some spiders are genuinely dangerous to humans. In principle, all spiders are venomous and should be treated with caution. Fortunately, most are not poisonous enough, or capable of injecting sufficient venom, to be truly harmful. Spider bites can be painful, but many spiders are too small, and their fangs too weak, even to penetrate human skin.

The real threat in the United States and Canada is posed by the so-called widow and brown spiders, which have a deservedly nasty reputation. Given the severity of their bites, it is not surprising that their venoms have been studied both chemically and pharmacologically.

The black widow, *Latrodectus mactans,* is found in many warm parts of the world. Related species occur in Canada and the southern United States. Chemically, black widow venom contains a mix of neurotoxic proteins, together with enzymes that digest tissue and physically open the way for the penetration of venom, thereby triggering a cascade of pain-inducing events. Additional small-molecule constituents potentiate the effect of the neurotoxins. The neurotoxins, of which a number have been characterized, share the same fundamental chemical structure, although they differ in ways that render them differentially active against vertebrate, insectan, or crustacean nervous systems. All act by causing a massive release of neurotransmitters, especially at the neuromuscular junction, and then a depletion of neurotransmitters, such that communication between neuron and muscle is compromised. Human fatalities are typically the result of respiratory failure. Early symptoms indicative of possible serious eventual complications may include muscle cramping, nausea and vomiting, as well as severe headache and anxiety. Victims may require hospitalization and close monitoring. An antivenin is available, but is used usually for treatment of extreme cases only, since it may itself induce side effects.

The brown spiders are a problem in the Americas, and one species, the brown recluse, or fiddleback, spider *(Loxosceles reclusa),* is particularly troublesome. Its bite may cause extensive ulceration and hemolysis, a condition termed necrotising arachnidism. *Loxosceles rufescens* in Australia and *L. laeta* in South America cause the same condition. The brown recluse, as implied by its name, is hardly aggressive. But it may seek shelter in shoes, pockets, or bedclothes, and as a result be inadvertently provoked. Its fangs are too small to penetrate clothing, with the result that bites tend to be restricted to areas of exposed skin. The venom of brown spiders differs chemically from that of many other known spider venoms in that it does not contain specific neurotoxins. Sphingomyelinase D, a major component of brown recluse venom, acts by hydrolyzing sphingomyelin, one of the four principal phospholipids in cell membranes. The enzyme may also induce the release of the tumor necrosis factor TNF-alpha, which could in turn lead to prostaglandin production, and consequently to the induction of pain, inflammation, and cell death at the site of the bite. Sphingomyelinase D is found only in the closely related *Loxosceles* and *Sicarius* spider genera and in some bacteria, including *Clostridium perfringens,* a common cause of food poisoning. Other components of brown recluse venom are hyaluronidase and several proteolytic enzymes, which could themselves contribute to the pathological action of the fluid. No antivenin to brown recluse venom is currently available in North America, and there is considerable controversy as to how to treat human victims.

A number of spiders are known worldwide that are also potently venomous. In Australia, the funnel web spiders are a serious threat to humans. The Sydney funnel web, *Atrax robustus,* is one of the most dangerous spiders known, its bite being potentially lethal in a matter of hours, unless the antivenin is administered. This large, stout-bodied, and aggressive spider has powerful fangs and large venom sacs. Primates are more sensitive to the venom than are other mammals, which may possess endogenous venom inhibitors. The mix of venom components in this spider is different in males and females, the male being the more dangerous, on account of its possession of robustotoxin, the lethal component of the venom. This neurotoxin interferes with the inactivation of neuronal sodium channels, causing neurons to be overly active. The exacerbated neuronal activity leads to a progression of symptoms, ranging from extreme pain at the site of the bite to eventual respiratory and circulatory failure. An antivenin has been in use since 1980, and there have been no known fatalities since then. From a biological standpoint, it is noteworthy that funnel web spider venom has a multiplicity of components, aside from the specific factor lethal to humans. From one Australian funnel web species as many as eight separate factors have been characterized that show specific anti-insectan activity. Given that insects are the principal prey of spiders, this should come as no surprise.

Because of their effect on neuronal systems, and in particular be-

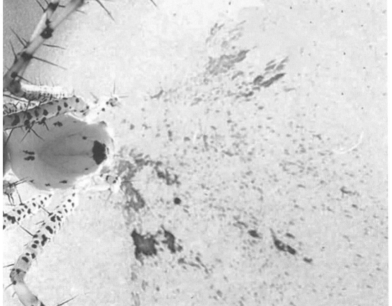

Peucetia viridans. Top: Feeding. Bottom: An individual that was persistently disturbed and caused to eject its venom in multiple directions. The spray patterns are rendered visible on a substrate of ground glass illuminated from beneath.

cause of their specificity of action vis-à-vis vertebrate or invertebrate target systems, spider venom components are currently the subject of increased study. It is hoped that this research will lead to the isolation of novel pharmacological agents of medicinal use and also, perhaps, anti-insectan toxins applicable in pest control.

It is typical for spiders to inject their venom, whether they are using it offensively or defensively. Until recently, no spiders had been known to eject their venom as a spray. But an exception has now been discovered. The green lynx spider, *Peucetia viridans,* when directly disturbed, as when it is seized by a leg with forceps, revolves the body so as to face the assaulting instrument, and discharges a jet of venom from the fangs. The fluid is ejected with considerable force and may range to a distance of 20 centimeters. Nothing is known about the defensive effectiveness of the spray, and it remains unknown whether the spider uses the spray only in response to direct assaults or also preemptively, as possibly when approached by a bird. So far as is known, lynx spiders also use their fangs in the usual fashion, for injection of venom into prey. Lynx spiders have good vision and are agile hunters. The females fasten their egg sacs to vegetation and guard the eggs, as well as the emerging spiderlings. Lynx spiders have been observed to feed on a range of insects, including among others bees, wasps, and stink bugs.

Lynx spiders do not pose a problem to humans. Though a few cases of human envenomation have been reported, the bites appear to cause no more than local pain and swelling and no tissue necrosis or systemic aftereffects. Little is known about the chemistry of the venom, but since the fluid is occasionally sprayed, it would be interesting to know whether it is fashioned in some special way to be active topically.

REFERENCES

Bettini, S., and M. Maroli. 1978. Venoms of Theridiidae, genus *Latrodectus:* systematics, distribution, and biology of species; chemistry, pharmacology, and mode of action of venom. In S. Bettini, ed., *Arthropod Venoms.* Berlin: Springer-Verlag.

Fink, L. S. 1984. Venom spitting by the green lynx spider, *Peucetia viridans* (Araneae, Oxyopidae). *Journal of Arachnology* 12:372–373.

Levi, H. W., and R. W. Levi. 1968. *A Guide to the Spiders and Their Kin.* New York: Golden Press.

Rash, L. D., and W. C. Hodgson. 2002. Pharmacology and biochemistry of spider venoms. *Toxicon* 40:225–254.

6

Class CHILOPODA
Order SCOLOPENDRIDA
Family Scolopendridae
Scolopendra heros
The giant Sonoran centipede

Scolopendra heros.

The body of a centipede consists of a head and trunk, the latter made up of a chain of similar segments, each bearing a pair of legs. There is no division into thorax or abdomen. The trunk is flexible, which makes it possible for centipedes to circumvent objects or to coil their bodies, as they typically do when they wrap themselves protectively around their eggs. The head is small and not always recognizable as such, because the front and rear of a centipede are often similar in appearance. To a naturalist intent on grasping a centipede this can be disconcerting, because picking up a centipede by the rear leaves the animal free to bite with the front. Predators are likely to be similarly confounded. Most dangerous are the centipedes of the order Scolopendrida. These include some veritable monsters, reaching lengths of nearly 30 centimeters and able to inflict serious bites. *Scolopendra heros* is one such monster. When exposed in its hiding place, this centipede tends first to move jerkily forward and backward, rather than directionally in headlong escape. Front and rear may be similarly colored, and in that case the uncertainty as to which end is which is compounded.

Scolopendrids as a rule have fewer than the 100 legs implied by their name. *Scolopendra heros,* with its 21–23 pairs, is nonetheless capable of great agility and speed.

Scolopendrids inject their venom with the fangs, two sharply pointed, pincer-like structures, situated immediately behind the head. Each fang has its own venom gland, equipped with an ejaculatory duct opening by way of a pore near the tip of the fang. The fangs are operated by powerful muscles that ensure not only that the fangs are driven into the victim but that the glands are simultaneously compressed and caused to squeeze out venom.

Scolopendrids are nocturnal, and feed mainly on insects, spiders, and earthworms, although they may also on occasion take toads, frogs, lizards, and mice. They are not generally a lethal threat to humans, and tend to bite only when directly provoked.

Surprisingly little is known about the chemistry of *Scolopendra* venom. The fluid is difficult to obtain in amounts sufficient for study, and degrades readily when separated or chemically purified. Studies to date indicate that *Scolopendra* venom contains a mix of compounds, including toxins that are tailored for specific targets, as well as other constituents that complement or further the toxic effects, such as proteases and biogenic amines. One *Scolopendra* toxin has multiple effects on the insect nervous system, essentially rendering the insect numb to sensory information and therefore incapable of escape. Another toxin appears to target the vertebrate autonomic

The head of a scolopendrid centipede, showing the sharply
pointed fangs, black at the tip.

nervous system, affecting heart rate, respiration, and smooth muscle
tone in a manner that could prove lethal to smaller vertebrates.

Some 2,500 species of centipedes are known worldwide. Most
are nocturnal, and all are predaceous, having fangs with associated
venom glands. They are divided into four orders. The Scolopendrida are mostly of tropical distribution and include the menacing
giant forms. The Scutigerida include the familiar long-legged *Scutigera coleoptrata,* a centipede commonly found in cellars in North
America. Although small in comparison with the Scolopendrida,
this centipede is nonetheless capable of inflicting painful bites, and
efforts to evict it, as when one encounters it in the bathtub or
shower stall, should not be undertaken by hand. On balance, however, the presence of *S. coleoptrata* in one's home is not without its
benefits, since the animal is an effective insectivore. The other centipede orders are the Lithobiida, or stone centipedes, largely confined
to the temperate zones of the world, and the Geophilida, or soil centipedes (see Chapter 7), widely distributed from the tropics to the
Arctic Circle. Stone centipedes are the most common centipedes in
North American forests. Soil centipedes are burrowing forms, living
in soil or decaying logs. Soil and stone centipedes prey mainly on
other, smaller arthropods.

REFERENCES

Minelli, A. 1978. Secretions of centipedes. In S. Bettini, ed., *Arthropod Venoms.* Berlin: Springer-Verlag.

Shelley, R. M. 2002. *A Synopsis of the North American Centipedes of the Order Scolopendromorpha (Chilopoda),* Memoir 5. Martinsville, Va.: Virginia Museum of Natural History.

7

Class **CHILOPODA**
Order **GEOPHILIDA**
Family Oryidae
Orphnaeus brasilianus
A geophilid centipede

A female *Orphnaeus brasilianus* guarding her eggs.

Fangs may be the universal weapon of centipedes, but some species have evolved additional defenses. The stone centipedes (order Lithobiida) have glands in the last two pairs of legs, from which they eject a sticky material that serves to entangle enemies. The soil centipedes (order Geophilida) add yet another, and very potent, defense. They have unicellular glands along the ventral surface of the body, opening through closely spaced microscopic pores. The secretion they

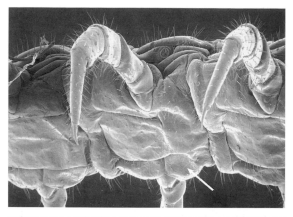

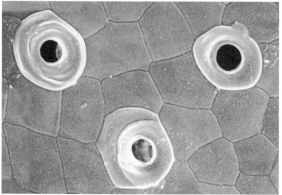

A ventrolateral view of a portion of the body of a geophilid centipede. The arrow in the top photo denotes the region of glandular pores, such as are shown enlarged in the bottom photo.

produce is sticky and has been shown in some species to be proteinaceous. Most remarkable, however, is that the secretion is cyanogenic, meaning that it generates hydrogen cyanide (HCN). Hydrogen cyanide is a nearly universal toxicant. It interferes with the fundamental respiratory processes of living cells by blocking the action of cytochrome oxidase enzymes that are essential to respiration.

Hydrogen cyanide is a gas at ordinary temperatures and therefore difficult to store as such. Geophilids solve the problem by storing the precursor molecules, mandelonitrile (12) and benzoyl cyanide (13). When the secretion is ejected, these compounds break down, the former into HCN plus benzaldehyde (14), the latter into HCN plus benzoic acid (15). One can imagine the dissociation of mandelonitrile and benzoyl cyanide to be catalyzed by special enzymes, brought into action somehow when the secretion is discharged. The mechanism might be analogous to that in hydrogen cyanide–liberating millipedes (see Chapter 9).

The secretion of geophilids is potently effective against ants and spiders, both undoubtedly natural enemies of centipedes. All constituents of the secretion, the sticky proteins as well as the accompanying noxious products, contribute to the action of the fluid. Hydrogen cyanide may well be the principal deterrent in the mixture, but benzaldehyde and benzoic acid can be expected to be active in their own right. Benzaldehyde, in fact, has been shown to be repellent to ants.

Geophilid females guard their eggs, and when challenged while doing so, may call their defenses into action. The picture at the beginning of this chapter shows a female *Orphnaeus brasilianus* coiled around her eggs in typical guarding posture. The female, when found, was concealed in an empty moth cocoon on Lignum Vitae Key in Florida, and was exposed for photographic purposes by cutting away some of the cocoon wall. When gently poked with forceps, she rotated her body so as to expose the ventral surface, and discharged copious amounts of her sticky cyanogenic secretion.

In some geophilid centipedes, the secretion is luminescent. Such is the case with the fluid discharged by *Geophilus vittatus,* which emits a faint blue-green glow for some seconds after emission. Whether the glow contributes to the defensive effectiveness of the secretion is unknown.

The Geophilida are of worldwide distribution, but are little known. The last comprehensive review of the order is over 50 years old. Some 1,000 species are estimated to exist.

Hydrogen cyanide production has evolved in a number of arthropod lineages, beside geophilid centipedes and polydesmid millipedes (see Chapter 9). It is known to occur, for instance, in certain beetle larvae, and in Lepidoptera (see Chapter 59).

The presumed mechanism of hydrogen cyanide (HCN) production in geophilid centipedes, by dissociation of mandelonitrile (12) and benzoyl cyanide (13). The breakdown products, aside from HCN, are respectively benzaldehyde (14) and benzoic acid (15).

REFERENCES

Jones, T. H., W. E. Conner, J. Meinwald, H. E. Eisner, and T. Eisner. 1976. Benzoyl cyanide and mandelonitrile in the cyanogenetic secretion of a centipede. *Journal of Chemical Ecology* 2:421–429.
Koch, A. 1927. Studien an Leuchtenden Tieren. I. Das Leuchten der Myriapoden. *Zeitschrift für Ökologie der Tiere* 8:241–270.

8

Class DIPLOPODA

Order SPIROBOLIDA

Family Floridobolidae

Floridobolus penneri

The Florida scrub millipede

Floridobolus penneri discharging secretion

Millipedes are an ancient group, dating back to Devonian times. About 10,000 contemporary species have been described, but hundreds more probably exist. Slow and sluggish despite their many legs, millipedes are furtive vegetarian scavengers, active primarily at night. Although few in kind, at least relative to the insects, millipedes are not a "defeated" group. They were succeeded by insects in evolution, but they were not displaced. Being agile predators, insects posed a formidable threat, but thanks to effective means of defense, millipedes were able to hold their own.

Anyone who has collected millipedes in the field knows that these animals commonly give off noxious fluids when disturbed. Malodorous and often irritating, the liquids stem from special glands, and are often discharged in copious quantity. The glands and their secretions have been studied in recent years, and much has been learned about the chemistry of the fluids and how millipedes use them in defense.

Although a diverse lot, millipedes have certain characteristics in common. Like centipedes, millipedes have a body made up of a head and a trunk, the latter consisting of a series of segments bearing the legs. Unlike centipedes, however, millipedes have two pairs of legs per segment. Each millipede segment is actually a fused pair of segments, which accounts for the relative inflexibility of the millipede's body. Centipedes scurry around obstacles, while millipedes tend to plow along in bulldozer fashion. Millipedes are, however, able to roll their bodies into a coil or flat spiral, or even into a sphere, and they avail themselves of this capacity when disturbed. Touch a millipede and chances are it will coil up. Disturb it some more, and the probability is high that it will emit defensive fluid. Some millipedes respond by coiling when merely breathed upon at close range (see Chapter 50). They seem to be programmed to take preemptive action before an attack is even initiated.

Floridobolus penneri is endemic to the so-called scrub habitat, characteristic of the highlands of central Florida. This habitat is rapidly disappearing as a consequence of human encroachment, and *F. penneri* itself is probably endangered. Although it is a member of one of the larger orders of millipedes, the Spirobolida, within that order *F. penneri* belongs to a family all its own, the Floridobolidae. Its disappearance would therefore leave a gap of some significance in the faunal listing of present-day millipedes. *F. penneri* has been little studied, except for its defenses. These are powerful, and of proven effectiveness against predators (although they will not help the animal withstand the loss of habitat).

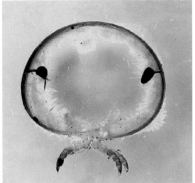

Left: The defensive glands of *Floridobolus penneri,* nestled amidst the body wall musculature. Right: A body segment of *F. penneri,* with all organs removed except the two defensive glands, which are replete with secretion.

F. penneri lives on sandy terrain and the animal leaves a conspicuous track when it crawls about during the night. Like other members of the order Spirobolida, it has glands distributed serially along the length of the body, one pair per segment, with openings visible as small pores on the flanks. Each gland is a spherical sac, embedded in the musculature of the body wall, and usually full to capacity with secretion. The sac bears a short exit duct, which leads to the pore and is ordinarily kept occluded by a valve. A special muscle operates this valve and is called into action when the millipede discharges its secretion. The sac itself lacks a muscular coating. Its compression is presumably effected indirectly by a rise in pressure in the body fluid surrounding the gland. The millipede exercises some level of control over the amount of secretion it discharges. A disturbance does not necessarily cause it to void fluid from all glands at once. As a rule, the millipede discharges first from the glands closest to the site of the disturbance, and then from the remaining glands if the assault persists. Chemically, *F. penneri's* secretion consists of a mixture of 1,4-benzoquinones (**16–21**).

Rules in nature have their exceptions, and this applies to the effectiveness of *F. penneri's* secretion. Although the fluid is repellent to most predators, there is one insect, the larva of a beetle of the family

16. 2-Methyl-1,4-benzoquinone   **17.** 2-Hydroxy-3-methyl-1,4-benzoquinone

18. 2-Methoxy-3-methyl-1,4-benzoquinone   **19.** 3-Methoxy-2,5-dimethyl-1,4-benzoquinone

20. 2,3-Dimethoxy-1,4-benzoquinone   **21.** 2,3-Dimethoxy-5-methyl-1,4-benzoquinone

Phengodidae, that preys on *F. penneri* as a matter of routine. *Phengodes lateicollis* shares *F. penneri*'s habitat. Its wormlike larva manages to kill *F. penneri* without risking exposure to the millipede's secretion. The larva overpowers *F. penneri* by coiling itself around the millipede's front end and injecting a hefty dose of gut fluid into the millipede's neck through perforations that it creates with its needle-sharp mandibles. The fluid paralyzes the millipede instantly, immobilizing all its muscles, including those that operate the opener valves of the defensive glands. The millipede is thus prevented from using its glands, and as it dies and its body tissues are liquefied by the injected fluid, the larva proceeds to imbibe the contents. Only the millipede's skeletal armor is eventually discarded, together with

A *Phengodes lateicollis* larva in the process of eating a *Floridobolus penneri.*

the defensive glands, which remain intact because they are lined with an indigestible membrane.

The order to which *F. penneri* belongs, the Spirobolida, includes some 10 families and 450 species. Spirobolids are most common and diverse in the tropics.

Benzoquinones are of such frequent occurrence in the defensive glands of arthropods that one cannot help wondering whether arthropods are for some biochemical reason predisposed to produce these chemicals. Among insects, benzoquinones are known from the defensive glands of earwigs, termites, cockroaches, grasshoppers, and beetles (for example, see Chapters 15, 16, 17, 21, 35, 49, 50, and 51). Among other arthropods, they are known from opilionids (Chapter 2) as well as millipedes of three orders (Spirobolida, Spirostreptida, Julida). Arthropods all produce benzoquinones as tanning agents when they molt. They use the chemicals to darken and harden the new exoskeleton every time they shed the old one. Producing benzoquinones is therefore part of the fundamental biochemical "know-how" of arthropods, a proficiency that may well have set the stage for the repeated evolution of quinone-secreting defensive glands within this group of animals.

REFERENCES

Attygalle, A. B., S.-C. Xu, J. Meinwald, and T. Eisner. 1993. Defensive secretion of the millipede *Floridobolus penneri*. *Journal of Natural Products* 56:1700–1706.

Eisner, T., D. Alsop, K. Hicks, and J. Meinwald. 1978. Defensive secretions of millipeds. In S. Bettini, ed., *Arthropod Venoms*. Berlin:
Springer-Verlag.

Eisner, T., M. Eisner, A. B. Attygalle, M. Deyrup, and J. Meinwald. 1998.
Rendering the inedible edible: circumvention of a millipede's chemical defense by a predaceous beetle larva (Phengodidae). *Proceedings of
the National Academy of Sciences USA* 95:1108–1113.

9

Class DIPLOPODA
Order POLYDESMIDA
Family Polydesmidae
Apheloria kleinpeteri
A polydesmid millipede

Apheloria kleinpeteri.

The Polydesmidae are remarkable in that they are cyanogenic: they produce hydrogen cyanide (HCN). Even before it became established that polydesmid millipedes were able to produce this powerfully toxic gas, naturalists had learned that these millipedes had a killing effect if confined in a closed jar with other life forms. Proof that the compound responsible for this lethal effect is HCN dates back to 1882.

Polydesmidae are medium-sized millipedes, about 5 centimeters or less in length. Their body, typically, bears lateral flanges, which project, one pair per segment, from above the legs. The defensive glands that produce HCN are housed within these flanges and open by way of small pores on the margins of the projections. The glands are bilateral, and are present in most body segments, with the result that the animal is well protected along both flanks. Predators tend to reject polydesmid millipedes.

The glands of polydesmids are especially designed for controlled release of HCN gas. The glands are two-chambered, each consisting of an inner storage chamber, or reservoir, and an outer reaction chamber. The reaction chamber is interposed between the reservoir and the outer opening of the gland.

The glands do not store HCN as such. Instead, they store cyanogenic compounds, substances that can be induced chemically to release HCN. The cyanogenic compounds are produced in the reservoir and are held there until forced through the reaction chamber at the moment of glandular discharge. The reaction chamber contains chemical factors, presumably enzymes, that catalyze the breakdown of the cyanogenic compounds, with the result that when the reservoir contents are squeezed through the reaction chamber, HCN release is initiated.

In polydesmids of the genus *Apheloria,* which have been studied

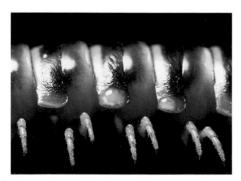

Left: Close-up lateral view of *Apheloria corrugata,* showing droplets of secretion emitted from two of the glands. Right: The toad *Bufo americanus* rejecting an *Apheloria corrugata* that it had just taken into the mouth.

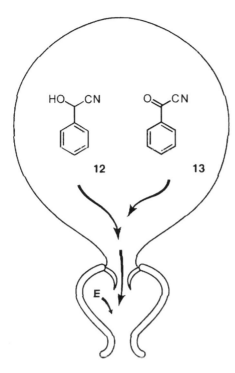

The mechanism of hydrogen cyanide (HCN) production in the gland of a polydesmid millipede. The principal cyanogenic compounds (12, mandelonitrile; 13, benzoyl cyanide) are stored in the gland's reservoir. The enzymes (E) in the reaction chamber promote the breakdown of 12 and 13, when these substances are forced through the reaction chamber at the moment of discharge. The breakdown products of 12 and 13 are, in addition to HCN, benzaldehyde (14) and benzoic acid (15), respectively.

in detail, the primary cyanogenic compound is mandelonitrile (12), a combination of HCN and benzaldehyde. A second cyanogenic compound, present in lesser amounts in the secretion, is benzoyl cyanide (13). Thus HCN production in this millipede involves the same chemical steps as it does in geophilid centipedes (see Chapter 7). The amount of HCN that can be released by polydesmid millipedes is substantial. An average-sized *Apheloria corrugata*, for instance, weighing about 1 gram, may produce as much as 600 micrograms of HCN. This is equivalent to 18 times the dose lethal to a pigeon, 6 times that lethal to a mouse, and 1/100 of that lethal to a human.

Other compounds released in the course of HCN production contribute to the effectiveness of a polydesmid's secretion. Thus the benzaldehyde liberated when mandelonitrile is broken down is itself repellent to ants. Further compounds, also isolated from polydesmid secretions, including a diversity of phenols, probably are deterrent as well.

Polydesmids often bear color markings (witness, for instance, *Apheloria kleinpeteri*), which, given that these animals are noxious, may be indicative of aposematism (warning coloration). Does it make sense for animals that are nocturnal to be aposematic? Actually, polydesmids are not strictly nocturnal. They are visible when migrating, which they have been known to do in the daytime, and also when they are dug up by diurnal predators.

The order Polydesmida, to which *Apheloria* belongs, includes 27 families. It is the largest order within the Diplopoda. The family Polydesmidae is itself large, comprising 200 species in 22 genera, all confined to the Northern Hemisphere.

REFERENCES

Duffey, S. S., and M. S. Blum. 1977. Phenol and guaiacol: biosynthesis, detoxication, and function in a polydesmid millipede, *Oxidus gracilis. Insect Biochemistry* 7:57–65.
Duffey, S. S., M. S. Blum, H. M. Fales, S. L. Evans, R. W. Roncadori, D. L. Tiemann, and Y. Nakagawa. 1977. Benzoyl cyanide and mandelonitrile benzoate in the defensive secretions of millipedes. *Journal of Chemical Ecology* 3:101–113.
Eisner, H. E., D. W. Alsop, and T. Eisner. 1967. Defense mechanisms of arthropods. XX. Quantitative assessment of hydrogen cyanide production in two species of millipedes. *Psyche* 74:107–117.
Eisner, T., D. Alsop, K. Hicks, and J. Meinwald. 1978. Defensive secretions of millipeds. In S. Bettini, ed., *Arthropod Venoms.* Berlin: Springer-Verlag.
Eisner, T., H. E. Eisner, J. J. Hurst, F. C. Kafatos, and J. Meinwald. 1963. Cyanogenic glandular apparatus of a millipede. *Science* 139:1218–1220.

Guldensteeden-Egeling, C. 1882. Über Bildung von Cyanwasserstoffsäure bei einem Myriapoden. *Pflügler's Archiv für die Gesamte Physiologie* 28:576–579.

Kuwahara, Y., H. Omura, and T. Tanabe. 2002. 2-Nitroethenylbenzenes as natural products in millipede defense secretions. *Naturwissenschaften* 89:308–310.

Taira, J., K. Nakamura, and Y. Higa. 2003. The identification of secretory compounds from the millipede, *Oxidus gracilis* C. L. Koch (Polydesmida: Paradoxosomatidae) and their variation in different habitats. *Applied Entomology and Zoology* 38:401–404.

10

Class **DIPLOPODA**
Order **POLYZONIIDA**
Family Polyzoniidae
Polyzonium rosalbum
A polyzoniid millipede

Polyzonium rosalbum.

The Polyzoniidae include only five genera of millipedes, mostly small and pale-colored. They are found in moist leaf litter and humus, and are sometimes locally abundant. *Polyzonium rosalbum* occurs in the eastern United States. Naturalists had noted that this millipede has an odor strongly reminiscent of camphor, stemming from a sticky whitish fluid that the animal emits when disturbed. Chemical analyses showed that this secretion contains two novel

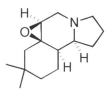

22. Polyzonimine   23. Nitropolyzonamine   24. Buzonamine

A *Formica* ant attacking (left) and then rejecting a *Polyzonium rosalbum.*

compounds, polyzonimine (**22**) and nitropolyzonamine (**23**), of which the former, although totally unrelated structurally to camphor, does indeed have a strong camphorous odor.

The secretions of millipedes, when put to the test, have consistently proved to be deterrent to ants. Since ants are probably among the principal enemies of millipedes, this should come as no surprise. Polyzonimine itself is formidably repellent to ants. If a glass capillary tube with polyzonimine is brought to within close proximity of ants that have been lured to a food source, the ants commence dispersing within a fraction of a second. *P. rosalbum* itself causes attacking ants to flee the moment it discharges its defensive fluid. The secretion, which oozes from glands that open along the millipede's flanks, sticks to the millipede's sides, and as a result protects the millipede against renewed assault as it leaves the site of action.

Another millipede of the order Polyzoniida, *Buzonium crassipes,* from California, has been shown to produce a novel defensive agent, buzonamine (**24**), which is, not surprisingly, also repellent to ants.

The dispersal of *Formica* ants in response to polyzonimine, presented in a glass capillary tube. The time course, beginning with t = 0 (top left), is 0.5 second (top right), 1.5 second (bottom left), and 2.5 second (bottom right).

REFERENCES

Meinwald, J., J. Smolanoff, A. T. McPhail, R. W. Miller, T. Eisner, and K. Hicks. 1975. Nitropolyzonamine: a spirocyclic nitro compound from the defensive glands of a milliped (*Polyzonium rosalbum*). *Tetrahedron Letters* 28:2367–2370.

Mori, K., and Y. Takagi. 2000. Enantioselective synthesis of polyzonimine and nitropolyzonamine, spirocyclic components in the defensive glands of a millipede, *Polyzonium rosalbum*. *Tetrahedron Letters* 41:6623–6625.

Smolanoff, J., A. F. Kluge, J. Meinwald, A. McPhail, R. W. Miller, K. Hicks, and T. Eisner. 1975. Polyzonimine: a novel terpenoid insect repellent produced by a milliped. *Science* 188:734–736.

Wood, F. W., J. H. Hanke, I. Kubo, J. A. Carrol, and P. Crews. 2000. Buzonamine, a new alkaloid from the defensive secretion of the millipede, *Buzonium crassipes*. *Biochemistry, Systematics, and Ecology* 28:305–312.

11

Class DIPLOPODA

Order GLOMERIDA

Family Glomeridae

Glomeris marginata

A pill millipede

Glomeris marginata, uncoiled and coiled.

The order Glomerida comprises millipedes that share one unusual feature, the ability to coil into a sphere or "pill." When coiled, such millipedes keep the appendages tucked away, to obvious defensive advantage. Glomerids are also chemically protected, by glands arranged two per segment, with openings along the mid-dorsal line. The best studied of these glomerids is a European species, *Glomeris marginata,* often found in abundance in forested habitats.

 G. marginata resorts to coiling as a first line of defense. If predators persist, the millipede then calls its glands into action. The glandular secretion is colorless and viscous, and is discharged as a series

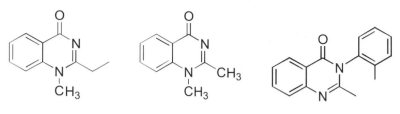

25. Homoglomerin   26. Glomerin   27. Methaqualone

of droplets that readily stick to alien surfaces. Ants are quickly immobilized by the secretion, which they spread over their bodies as they attempt to wipe the fluid away with the legs. The secretion is bitter-tasting, on account of the two principal components, homoglomerin (**25**) and glomerin (**26**). These compounds belong to a category of organic substances called quinazolinones, not known from any other animals.

Experiments with spiders showed that glomerin and homoglomerin are paralyzing agents. Wolf spiders, for instance, tend to reject *G. marginata* on the basis of taste, but they sometimes do so only after having killed the millipede and ingested some of the secretion. When this is the case, the spiders may become totally paralyzed, sometimes for days on end, a condition that under natural circumstances would doubtless prove fatal. One may wonder how this benefits the millipede, given that it may itself lose its life in the encounter. In fact, it is the genetic kin of the victimized millipede that may profit. Millipedes lack the means for quick dispersal and can therefore be expected to coexist with close relatives. To lose an occasional individual from such a group to the process of predator-culling makes sense, since such a loss is for the "common good." One would be hard put to envision how a defense mechanism might evolve that depends on loss of individuals to predation, unless the survivors who benefit from the loss are genetically related to the individuals killed by the predator.

The *G. marginata* secretion is toxic to other predators as well. Consumption of *G. marginata* can lead to partial paralysis in mice, to slowed reactions in birds, and to vomiting in toads. Interestingly, there is a quinazolinone closely related to glomerin and homo-

Top: *Glomeris marginata,* discharging from its eight
pairs of defensive glands. Bottom: A wolf spider,
paralyzed from having ingested some of a *G.
marginata*'s secretion in the course of an attack
upon the millipede, being assaulted by ants.

glomerin, the synthetic substance called methaqualone, or Quaa-
lude (27), which because of its sedative action has been put to use as
a drug by humans.

Another order of millipedes, the Spherotheriida, closely related to
the Glomerida, includes large species that also coil into a sphere.
Endowed with a tough shell, these millipedes lack the chemical defenses of glomerids. Although seemingly impregnable and doubtless

rejected by many predators, spherotheriids are eaten by mongooses in South Africa. These predators grasp the coiled millipedes in the paws, and after positioning themselves with their backside close to a solid structure, such as a tree trunk, hurl the millipedes backward between the legs and smash them to bits. They then feast on the pieces.

REFERENCES

Carrel, J. E., and T. Eisner. 1984. Spider sedation induced by defensive chemicals of milliped prey. *Proceedings of the National Academy of Sciences USA* 81:806–810.

Eisner, T., and J. A. Davis. 1967. Mongoose throwing and smashing millipedes. *Science* 155:577–579.

Eisner, T., D. Alsop, K. Hicks, and J. Meinwald. 1978. Defensive secretions of millipeds. In S. Bettini, ed., *Arthropod Venoms.* Berlin: Springer-Verlag.

Meinwald, Y. C., J. Meinwald, and T. Eisner. 1966. 1,2-Dialkyl-4 (3H)-quinazolinones in the defensive secretion of a millipede (*Glomeris marginata*). *Science* 154:390–391.

Schildknecht, H., U. Maschwitz, and W. F. Wenneis. 1967. Neue Stoffe aus dem Wehrsekret der Diplopodengattung *Glomeris.* Über Arthropoden-Abwehrstoffe. XXIV. *Naturwissenschaften* 54:196–197.

12

Class DIPLOPODA

Order POLYXENIDA

Family Polyxenidae

Polyxenus fasciculatus

A bristle millipede

Polyxenus fasciculatus.

The Polyxenidae, known as bristle millipedes, are minute, highly aberrant diplopods, so aberrant in fact that it has been proposed that they deserve separate class status within the Arthropoda.

Only a few millimeters long, polyxenids have a body composed of only 11 segments. They are elaborately ornate in appearance due to a covering of modified bristles, neatly arranged in tufts and rows. Unlike millipedes generally, they have a skeletal shell devoid of calcium salts. There are some 50 described species of Polyxenidae, mostly from the Northern Hemisphere. They are partial to dry situations, and may occur locally in abundance. In central Florida they are sometimes found in quantity under the flaky bark of pine trees (*Pinus elliotti*).

The defense of polyxenids is purely mechanical, and involves the use of a remarkable tuft of bristles that projects from the rear of the body. The tuft actually consists of a pair of tufts, ordinarily held so closely appressed that they appear to form a single structure. The bristles of the tufts are loosely anchored and they bear what can best be described as a multi-pronged grappling hook at the tip. When a prowling ant comes in contact with a polyxenid, the millipede splays the tufts and brushes them against the ant. The millipede's response is so quick, and the splaying of the tufts so brief, that it is easy to miss the action. But the effect on the ant leaves no doubt that something dramatic has taken place. From one moment to the next, what was a vigorous, agile antagonist is reduced to a grotesquely sprawled, virtually immobilized entity, of no further threat to the millipede. When the tufts touch the ant, individual bristles take hold of the hairs (setae) projecting from the ant's surface. The bristles detach from the tufts, and thanks to the barbs along their shafts, interlock. The result is the formation of a network of bristles that tie together the appendages of the ant. As the ant attempts to wipe itself clean,

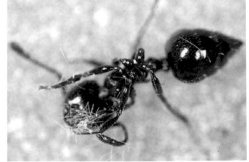

Left: An ant, moments after attacking a *Polyxenus fasciculatus,* immobilized by the bristles of the millipede. Right: Another ant, similarly immobilized, showing the tangle of bristles that has tied together the legs, antennae, and mouthparts.

Left: The terminal tufts of *Polyxenus fasciculatus*. Two bristles can be seen projecting free, their terminal hooks clearly discernible. Right top: The grappling hook at the tip of a *P. fasciculatus* bristle. Right bottom: A bristle that has grasped a hair (seta) on the surface of an ant.

using whichever legs might still be unencumbered, it succeeds only in spreading the bristles and increasing its entanglement. Polyxenids molt as adults, and every time they do they reconstitute their tufts. Bristle losses that they incur in the course of defensive action are therefore not permanent.

Though polyxenids are the only millipedes known to use detachable bristles for defense, they are not the only arthropods to derive protection from detachable integumental structures. Moths, for instance, are prevented from sticking to spider webs by their covering of scales (see Chapter 62). When they fly into a web they may simply flutter free, impoverished only by the loss of those scales that are torn off by contact with the viscid threads of the web. Insects covered with powdered wax, such as the so-called whiteflies (Hemiptera of the family Aleyrodidae) (see Chapter 30) may also fail to stick to spider webs. Remarkably similar to polyxenids are certain beetle larvae of the family Dermestidae. These too have a tuft of detachable bristles projecting from the rear, which they use in polyxenid-fashion to entangle enemies. Their bristles have terminal grappling hooks, very similar to those of polyxenids.

REFERENCES

Eisner, T., R. Alsop, and G. Ettershank. 1964. Adhesiveness of spider silk. *Science* 146:1058–1061.

Eisner, T., M. Eisner, and M. Deyrup. 1996. Millipede defense: use of detachable bristles to entangle ants. *Proceedings of the National Academy of Sciences USA* 93:10848–10851.

Ma, M., W. E. Burkholder, and S. D. Carlson. 1978. Supra-anal organ: a defensive mechanism of the furniture carpet beetle, *Anthrenus flavipes* (Coleoptera: Dermestidae). *Annals of the Entomological Society of America* 71:718–723.

Nutting, W. L., and H. G. Spangler. 1969. The hastate setae of certain dermestid larvae: an entangling defense mechanism. *Annals of the Entomological Society of America* 62:763–769.

13

Class INSECTA
Order DYCTIOPTERA
Family Blattidae
Eurycotis floridana
The Florida woods cockroach

A female *Eurycotis floridana* with an egg case.

Insects, it has been said, reign supreme on earth, and cockroaches are the elite. Cockroaches thrive the world over, even where humans have destroyed much of the natural environment. They share our resources wherever we live, and they are the only genuine majority in many of our cities.

Defenses are numerous and varied in cockroaches, and may involve protection of the self or of offspring. Some cockroaches coil into a sphere when attacked, others hug the ground so they cannot

28. *(E)*-2-Hexenal   **29.** Ethyl acrolein

be grasped, and still others derive protection from defensive glands. *Eurycotis floridana* relies on a gland for protection. The gland is a large pouch, situated in the abdomen of adult males and females, and opening as a ventral slit between the sixth and seventh segments. The gland is present in rudimentary form in the immatures (the nymphs) but appears to be functional as a dischargeable sac in the adults only.

E. floridana is a large wingless cockroach, abundant in the wild in Florida, where it occurs in stumps, under the bark of dead trees, and in ground litter. At night it commonly ventures outside such hiding places.

The secretion produced by *E. floridana*'s abdominal gland contains some 40 components, including aldehydes, alcohols, carboxylic acids, lactones, and an ether. The primary component by far is *(E)*-2-hexenal (**28**), a powerful odorant, known also from a variety of other natural sources. It is produced, for instance, by many stink bugs (see Chapter 22) and also, at low concentrations, by many plants. Because of its frequent occurrence in plants, it is sometimes called "leaf aldehyde." Stink bugs and *E. floridana* adults produce the aldehyde at high concentrations.

E. floridana discharges readily when roughly handled. It ejects its secretion with considerable force, to distances of 10 to 20 centimeters or more. The fluid is sharply irritating to humans, both to the exposed surfaces of the face and upon inhalation. Getting sprayed in the eyes is excruciatingly painful, certainly to us, but to vertebrate predators undoubtedly as well. Indeed, the secretion has been shown to offer *E. floridana* protection against mice, birds, lizards, and frogs. It is an irritant to insects as well.

Because *E. floridana* nymphs lack secretion, they are potentially vulnerable. However, *E. floridana* are flightless, with the result that the older nymphs closely resemble the wingless adults in appear-

Top left: An *Eurycotis floridana*, tethered and placed on a sheet of indicator paper, showing the spray pattern of two discharges elicted by stimulating the hindlegs one at a time. Top right: A wolf spider feeding on an *E. floridana* adult. Wolf spiders are not always repelled by this insect's secretion. Bottom left: A female *E. floridana*, digging a trench for deposition of her egg case. Bottom right: A female *E. floridana*, having buried her egg case, is smoothing over the site to eliminate all evidence of her activity.

ance. Older nymphs could therefore be confused with adults by predators, and shunned as a result. In a sense, therefore, older *E. floridana* nymphs could be thought of as being mimics of the adults.

Some cockroaches related to *E. floridana* possess the same kind of defensive gland but do not discharge the same products. Australian species of the genus *Platyzosteria* are remarkable in that they produce ethyl acrolein (**29**), a veritable tear gas, at a concentration upward of 90%. It is not surprising that such secretions repel predators. What is surprising is that there should exist animals able to produce and tolerate such toxicants.

Eurycotis floridana egg cases. The case on the right, encrusted with sand, has been dug up from its burial site. The one on the left has been removed from the rear of a female before it was buried.

In addition to being protected as adults, *E. floridana* are well protected as eggs. The adult female, in fact, has an elaborate procedure for concealing the eggs by burying them. Cockroaches typically produce their eggs in packages, the oothecae, which often project from the rear of the female for some time before they are laid. Some cockroaches shelter their oothecae by incubating them internally, thus giving birth to living young. *E. floridana* females bury their egg cases in the sand. The behavior is intricate, and involves investment of time and resources on the part of the female. When ready, the female digs a trench by using her head as a shovel. She scoops back load upon load of sand, and when the trench is large enough, she lines it with copious amounts of saliva so that it will hold its shape. She then straddles the opening, and flexes her abdomen forward so as to slide the egg case into the hole. Finally, she scoops up more sand, wets it with saliva, and uses the mixture to cover up the hole. Before walking off, she takes great care to smooth over the entire site, eliminating all evidence of her activities. The saliva eventually dries, and then the egg case is entirely covered by a hardened crust of sand. The crust shields the egg case from view, and prevents it from being exposed when a predator digs in the sand in search of food. The amount of saliva used by the female to encase her ootheca is substantial. In a female about to oviposit the salivary glands are so distended that they take up much of her body cavity. After oviposition, the salivary glands are almost totally drained.

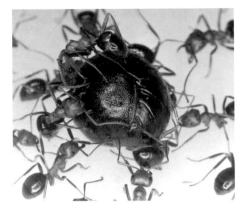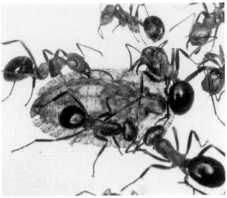

Left: A cockroach, *Perisphaerus semilunatus,* coiling into a sphere in response to an ant attack. Right: A cockroach (probably *Capucina* species) hugging the substrate when attacked.

No defense is perfect. There is a wasp that parasitizes *E. floridana* by laying its own eggs into the cockroaches' oothecae. It locates *E. floridana* females and oviposits into the oothecae while these are still projecting from the rear of the female.

The order Dyctioptera includes the cockroaches and the mantids. The cockroaches constitute 5 families with some 3,700 species. The family Blattidae, which also includes *Periplaneta* (see Chapter 14) and *Deropeltis* (see Chapter 15) contains over 500 species in 44 genera.

REFERENCES

Brossut, R. 1983. Allomonal secretions of cockroaches. *Journal of Chemical Ecology* 9:143–158.

Brossut, R., and L. Sreng. 1980. Ultrastructure comparée des glandes exocrines abdominales des Blattaria (Dyctioptera). *International Journal of Morphology and Embryology* 9:199–213.

Eisner, T., and M. Eisner. 2002. Coiling into a sphere: defensive behavior of a trash-carrying chrysopid larva (*Leucochrysa (Nodita) pavida*). *Entomological News* 113:6–10.

Farine, J-P., C. Everaerts, D. Abed, and R. Brossut. 2000. Production, regeneration, and biochemical precursors of the major components of the defensive secretion of *Eurycotis floridana* (Dictyoptera, Polyzosteriinae). *Insect Biochemistry and Molecular Biology* 30:601–608.

Farine, J.-P., C. Everaerts, J.-L. LeQuere, E. Semon, R. Henry, and R. Brossut. 1997. The defensive secretion of *Eurycotis floridana* (Dictyoptera, Blattidae, Polyzosteriinae): chemical identification and evidence of an alarm function. *Insect Biochemistry and Molecular Biology* 27:577–586.

Maschwitz, U., and Y. P. Tho. 1978. Phenols as defensive secretion in a Malayan cockroach, *Archiblatta hoereni* Vollenhoven. *Journal of Chemical Ecology* 4:375–381.

McKittrick, F. A. 1964. *Evolutionary Studies of Cockroaches,* Memoir 389. Ithaca: Cornell University Agricultural Experiment Station.

Roth, L. M., and D. W. Alsop. 1978. Toxins of Blattaria. In S. Bettini, ed., *Arthropod Venoms.* Berlin: Springer-Verlag.

Roth, L. M., and E. R. Willis. 1954. *Anastatus floridanus* (Hymenoptera: Eupelmidea), a new parasite on the eggs of the cockroach *Eurycotis floridana. Transactions of the American Entomological Society* 80:29–41.

Stay, B. 1957. The sternal scent gland of *Eurycotis floridana* (Blattaria: Blattidae). *Annals of the Entomological Society of America* 50:514–519.

Trumbull, M. W., and N. J. Fashing. 2002. Efficacy of the ventral abdominal secretion of the cockroach *Eurycotis floridana* (Blattaria: Blattidae) as a defense allomone. *Journal of Insect Behavior* 15:369–384.

Wallbank, B. E., and D. F. Waterhouse. 1970. The defensive secretion of *Polyzosteria* and related cockroaches. *Journal of Insect Physiology* 16:2081–2096.

Waterhouse, D. F., and B. E. Wallbank. 1967. 2-Methylene butanal and related compounds in the defensive scent of *Platyzosteria* cockroaches (Blattidae: Polyzosteriinae). *Journal of Insect Physiology* 13:1657–1669.

14

Class INSECTA
Order DYCTIOPTERA
Family Blattidae
Periplaneta australasiae
The Australian cockroach

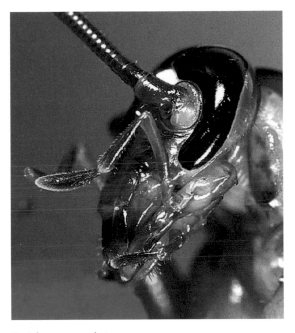

Periplaneta australasiae.

Crayfish, which are not fish at all, but freshwater relatives of the lobster, have a remarkable ability. They can detect light with the rear of the abdomen. Thus a crayfish has the capacity to sense when its back end is sticking out from beneath its hiding place. It does not have posterior eyes for the purpose, but instead senses light with the most posterior abdominal ganglion. The ganglion is close to the body surface under mostly translucent tissue, and responds by sending out nerve impulses when stimulated by light. For an animal that seeks concealment much of the time, but may be in quarters too cramped to check whether the rear is projecting free, having a rear that does its own light-metering is obviously advantageous.

Cockroaches, too, crave concealment. Anyone who has shared a kitchen with cockroaches knows that they seek shelter by day and that they are driven to flee for cover at night if a light is turned on. In nature, cockroaches also remain in hiding by day. Unlike crayfish, however, they are blind except up front, and therefore unable to check on their rear's light exposure. In cockroaches, however, protection of the rear may be effected by chemical means. In the immatures (the nymphs) of some species, including *Periplaneta australasiae, P. fuliginosa, P. japonica,* and *Blatta orientalis,* the tip of the abdomen bears a coating of slimy secretion, proteinaceous in nature, that shields the rear against assault. The sticky material coats the sixth and seventh abdominal tergites, as well as the upper surface of the two tapering sensory appendages (the cerci) that project from the rear. The secretion acts as a mechanical encumbering agent in fending off attacks by ants, and probably also spiders and other arthropods. It could therefore protect these roaches against some real dangers. The cerci, incidentally, are covered with hairs that are highly responsive to air motions, and capable, therefore, of forewarning a cockroach against impending assault from the rear.

Analytical work has failed to reveal the presence of repellents or otherwise active substances in the abdominal slime of cockroaches. The slime appears therefore to act strictly by virtue of its stickiness. Interestingly, some nymphs, notably those of *P. fuliginosa* and *P. japonica,* possess the ability to splash droplets of the secretion over some distance, and they do so when attacked by ants. Ants that are targeted by the fluid have been observed to be instantly incapacitated.

One species of cockroach, *Megaloblatta blaberoides,* also known to produce abdominal slime in the nymphal stage, is brightly colored and in the habit, as a nymph, of emitting a sound when disturbed. It

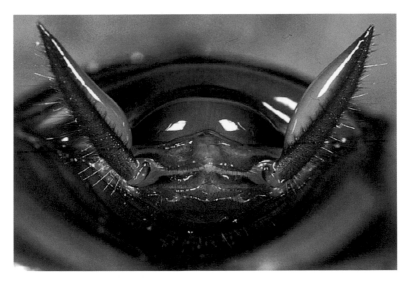

Rear-end view of a *Periplaneta australasiae,* showing the coating of slimy secretion. The two projecting spine-shaped structures are the cerci.

produces the sound by using specialized structures on the abdomen, a pair of "files" and "scrapers," which are rubbed against one another. It has been suggested that both the vivid color and the acoustic signal are aposematic, forewarning predators of the protected status of their intended target. The bright color could warn in the daytime, and the sound could warn at night.

Cockroaches are an ancient group, consisting of some 3,700 described species. Chemical defenses occur in some (see Chapters 13, 15, 16) but not all species.

REFERENCES

Bruno, M. S., and D. Kennedy. 1962. Spectral sensitivity of photoreceptor neurons in the 6th ganglion of the crayfish. *Comparative Biochemistry and Physiology* 6:41–46.

Ichinose, T., and K. Zennyoji. 1980. Defensive behavior of the cockroaches *Periplaneta fuliginosa* and *Periplaneta japonica* (Orthoptera Blattidae) in relation to their viscous secretion. *Applied Entomology and Zoology* 15:400–408.

Plattner, H., M. Salpeter, J. E. Carrel, and T. Eisner. 1972. Struktur und Funktion des Drüsenepithels der postabdominalen Tergite von *Blatta orientalis. Zeitschrift für Zellforschung* 125:45–87.

Roeder, K. D. 1967. *Nerve Cells and Insect Behavior.* Cambridge, Mass.: Harvard University Press.

Roth, L. M., and W. H. Stahl. 1956. Tergal and cercal secretion of *Blatta orientalis* L. *Science* 123:798–799.

Schal, C., J. Fraser, and W. J. Bell. 1982. Disturbance stridulation and chemical defense in nymphs of the tropical cockroach *Megaloblatta blaberoides. Journal of Insect Physiology* 28:541–552.

15

Class **INSECTA**

Order **DYCTIOPTERA**

Family Blattidae

Deropeltis wahlbergi

A blattid cockroach

A *Deropeltis wahlbergi* male in the defensive posture assumed when discharging.

D*eropeltis wahlbergi* is a large South African cockroach with a pair of sizable defensive glands positioned side by side in the abdomen, and opening dorsally by way of a shared slit between abdominal segments 5 and 6. The nymphs (the immatures) have no difficulty expelling their secretion, since the abdomen is not covered by wings. The adult female is also wingless and therefore similarly unencumbered. A *D. wahlbergi* discharges its secretion forcibly, in the form of

Top: A female *Deropeltis wahlbergi* in defensive posture. Bottom: Another female *D. wahlbergi*, tethered and placed on a sheet of indicator paper, showing the spray pattern produced in response to pinching of the left foreleg. The female rotated the abdomen while discharging; hence the curved spray pattern.

jets. Nymphs and females assume a rigid stance just before ejection, in which they tighten the legs and arch the back. By swinging the abdomen around to one side or the other, nymphs and females can exercise some control over the direction of their ejections. The adult male, which has fully developed wings, rotates them upward and out of the way when ejecting. It, too, arches its back when discharg-

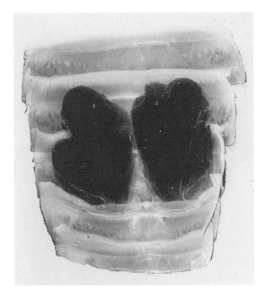

The defensive glands of a female *Deropeltis wahlbergi* stained so as to be visible through the abdominal wall.

30. 1,4-Benzoquinone

16. 2-Methyl-1,4-benzoquinone

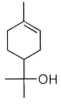

31. α-Terpineol

32. *(E)*-2-Dodecenal

ing, and directs its ejections in the same manner as the nymphs and females.

The secretion of *D. wahlbergi* contains 1,4-benzoquinone (**30**), and 2-methyl-1,4-benzoquinone (**16**), as well as the isoprenoid α-terpineol (**31**). Though it is not always the case that quinonoid secretions contain nonquinonoid constituents, *D. wahlbergi* is certainly not unique in this regard. The millipede *Rhinocricus insulatus,* for instance, secretes a mixture of 2-methyl-1,4-benzoquinone (**16**) and *(E)-2-dodecenal* (**32**). Both α-terpineol and *(E)-2-dodecenal* are deterrents in their own right, and can be expected to contribute to the defensive efficacy of their respective secretions.

REFERENCES

Roth, L. M., and D. W. Alsop. 1978. Toxins of Blattaria. In S. Bettini, ed., *Arthropod Venoms.* Berlin: Springer-Verlag.

Wheeler, J. W., J. Meinwald, J. J. Hurst, and T. Eisner. 1964. Trans-2-dodecenal and 2-methyl-l,4-quinone produced by a millipede. *Science* 144:540–541.

16

Class INSECTA
Order DYCTIOPTERA
Family Blaberidae
Diploptera punctata
The Pacific beetle cockroach

Diploptera punctata.

D*iploptera punctata* is a viviparous cockroach from the Pacific Islands. It has two defensive glands, situated in the abdomen, just behind the junction with the thorax. Individuals at all stages of development, the adults as well as the immatures, have the glands, which secrete a mixture of 1,4-benzoquinones (**16, 30, 33**). What is un-

16. 2-Methyl-1,4-benzoquinone   30. 1,4-Benzoquinone   33. 2-Ethyl-1,4-benzoquinone

usual about the glands is that they are discharged by use of respiratory air. Breathing in insects is effected by use of a series of internal tubes, called tracheae, that branch throughout the body. The tracheae open to the outside by way of special valvular orifices called spiracles, and within the body branch and rebranch to form extremely fine tubules, small enough to reach individual cells. Oxygen is thus conveyed directly to the tissues of an insect, rather than indirectly, in combination with respiratory pigments borne by the blood, as in so many other animals.

The *D. punctata* defensive glands consist of a pair of pouches, each one opening into a major tracheal tube just at the point where that tube connects to a spiracle. The glands are thus enabled to expel their contents through the spiracle. Force for the ejection is provided by bursts of air conveyed into each pouch through a narrow coiled duct, linked internally to the tracheal air sacs.

The secretion is ejected as a fine mist. The animal does not aim its discharges, although it is capable of ejecting from only one gland at a time. If disturbed unilaterally, it discharges exclusively from the gland of the side stimulated. Individual emissions are sufficiently dispersed to provide protection for an entire flank of the animal. Ants are repelled by the discharges. Individuals from which the glands have been surgically removed are overwhelmed by ants.

When *D. punctata* molts, it sheds the lining of the glands together with the exoskeleton. The gland contents are lost with the lining, with the result that *D. punctata* are vulnerable for a period of hours after molting, until their secretory supply has been replenished. The adult female has a special strategy for seeking protection during this time. Immediately after molting, while she is still light-

Top: Two *Diploptera punctata,* the one on the left with glands intact, the one on the right with glands removed, have been tethered, placed on indicator paper, and subjected to an attack by ants. The cockroach on the left repelled ants with glandular discharges. The glandless individual on the right is under persistent attack. Bottom: Two *D. punctata* mating. The male has discharged its defensive secretion in response to disturbance, providing protection for himself and his mate; the female, newly molted, is incapable of discharging.

colored and her exoskeleton has not fully hardened, she is attractive and responsive to courting males. Though she herself is defenseless during copulation, the male is not, with the result that the female may herself derive protection from the male's discharges, should the pair come under attack. The male, while copulating, sprays only if he himself is directly stimulated in the course of the assault. Molestation of the female alone does not cause him to spray.

Female *D. punctata* incubate their eggs internally. They have a true uterus, in the sense that the embryos within are provided with nutrients as they develop. Other cockroaches that incubate their eggs usually provide them with water only.

REFERENCES

Eisner, T. 1958. Spray mechanism of the cockroach *Diploptera punctata*. *Science* 128:148–149.

Roth, L. M., and D. W. Alsop. 1978. Toxins of Blattaria. In S. Bettini, ed., *Arthropod Venoms*. Berlin: Springer-Verlag.

Roth, L. M., and B. Stay. 1957. The occurrence of para-quinones in some arthropods, with emphasis on the quinone-secreting tracheal glands of *Diploptera punctata* (Blattaria). *Journal of Insect Physiology* 1:305–318.

Wyttenbach, R., and T. Eisner. 2001. Use of defensive glands during mating in a cockroach (*Diploptera punctata*). *Chemoecology* 11:25–28.

17

Class INSECTA

Order DERMAPTERA

Family Forficulidae

Doru taeniatum

An earwig

Doru taeniatum.

There are major orders of insects, such as the Coleoptera (the bee-tles), with its 300,000 species, and there are minor orders. With only 1,500 species, the earwigs (Dermaptera) are decidedly one of the latter. Earwigs are interesting, however, and not nearly well enough known. For young entomologists eager to explore uncharted territory, earwigs offer the promise of adventure.

Earwigs are slender insects, quick on their feet, and active mostly at night. They feed on plant and animal matter, and in the daytime are mostly in hiding. Their body is flexibly built, with loosely articu-

A tethered earwig *(Forficula auricularia)* defending itself
by use of its pincers. The animal is being pinched with
forceps on a foreleg (center) and on an antenna (bottom).

lated thoracic segments, and a long abdomen able to bend and ro-
tate in most directions. Characteristically, earwigs bear a pair of pincers at the end of the abdomen, which they use both as an aid in
prey capture and for defense. Pick up an earwig by hand, and
chances are it will attempt immediately to use its pincers in retalia-
tion. Restrict the attack by holding one of the earwig's appendages
with forceps, and it will flex its abdomen and attempt to grasp the
instrument and inflict a pinch.

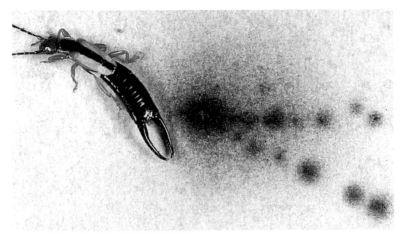

Top: The spray pattern (on indicator paper) of a *Doru taeniatum* that was stimulated by pinching the abdomen with forceps. Bottom: An inside view of the dorsal skeletal plate of the fourth abdominal segment of *D. taeniatum,* with glands attached. The glands are replete with yellow crystalline quinones.

In real life, the duel between an earwig and its attacker has a chemical dimension. Most earwigs have a pair of abdominal defensive glands, or even two pairs of such glands, opening dorsally by way of small pores, near the base of the abdomen. The glands are compressible and their contents are discharged as a spray. They are strategically located, in that their openings keep their aim at the pin-

16. 2-Methyl-1,4-benzoquinone

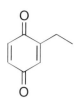

33. 2-Ethyl-1,4-benzoquinone

4. 2,3-Dimethyl-1,4-benzoquinone

34. Pentadecane

cers, no matter whether the abdomen is flexed right, left, or forward. The result is that the animal is able at all times to use its mechanical and chemical weapons in combination. Whatever it pinches with its pincers can also be hit by the spray.

Only two species of earwig have been investigated chemically, and the glands of both were shown to produce 1,4-benzoquinones. In the European earwig, *Forficula auricularia,* a species that has become established in North America, the glands (two pairs) store a dilute solution of the quinones (**16, 33**). In *Doru taeniatum,* the glands (one pair) are replete with a crystalline paste of quinones (**4, 16**). The individual crystals are too large to fit through the tiny outer openings of the glands. The result is that *D. taeniatum* discharges only the saturated solvent in which the quinones are stored—a mixture of water and the hydrocarbon pentadecane (**34**)—thereby economizing on the use of the quinones themselves. After a discharge, *D. taeniatum* needs only to replenish the expended solvent, which becomes saturated with quinones as it builds up in the glands and is mixed with the quinone crystals. Only moderate amounts of the quinones dissolve in the solvent, but such amounts suffice for defensive purposes, since benzoquinones are highly repellent at very low concentrations.

REFERENCES

Eisner, T. 1960. Defense mechanisms of arthropods. II. The chemical and mechanical weapons of an earwig. *Psyche* 67:62–70.

Eisner, T., C. Rossini, and M. Eisner. 2000. Chemical defense of an earwig (*Doru taeniatum*). *Chemoecology* 10:81–87.

Peschke, K., and T. Eisner. 1987. Defensive secretion of a beetle (*Blaps mucronata*): physical and chemical determinants of effectiveness. *Journal of Comparative Physiology* 161:377–388.

Vosseler, J. 1890. Die Stinkdrüsen der Forficuliden. *Archiv für Mikroskopische Anatomie* 36:565–578.

18

Class INSECTA
Order ISOPTERA
Family Termitidae
Nasutitermes exitiosus
A termite

Nasutitermes exitiosus, soldiers and a worker (at the bottom, center).

The year 1956 saw the addition of a useful term to the biological vocabulary. There had been growing awareness that communicative signals between organisms were often chemical in nature, rather than strictly visual or acoustic, and such signals were in need of a name. The term coined for the purpose, pheromone, is used to designate substances secreted to the outside by an organism that are capable of inducing a behavioral or physiological response in another organism of the same species. Pheromones stood defined as social chemical messengers, and they were an important category of substances.

Pheromones are ubiquitous in nature. All types of organsims, including animals, plants, and microorganisms, make use of them. In animals they may play a role in sex attraction and courtship, territorial marking, establishment of dominance hierarchies, and brood care. They are particularly important in social insects, which may depend on pheromones for nestmate recognition, regulation of caste ratios, recruitment to food sources, and advertisement of alarm. Termites in particular are dependent on pheromones, since they are virtually blind and are therefore unable to make use of visual signals.

All social insects use pheromones to communicate alarm. Honey bees, when they sting, leave the sting apparatus implanted in the victim. The apparatus is the source of a pheromone that alerts other bees to the site of "trouble," causing them to attend to the emergency (see Chapter 69). Ants, too, produce alarm pheromones. They have special glands for the purpose, from which they discharge when they are in trouble, causing nestmates to rally to the site (see Chapter 68). In termites, alarm pheromones are often produced by a special soldier caste. Such is the case in *Nasutitermes exitiosus.*

N. exitiosus is an Australian nasutitermite, a member of a widely distributed group of species (family Termitidae, subfamily Nasutitermitinae), whose soldiers, the nasutes, have a characteristically pointed snout, or rostrum, on the front of the head. The head capsule contains a large gland, which produces a sticky, odorous secretion that the solider discharges through the rostrum when disturbed. The odorous components of the secretion are simple monoterpenes, including α-pinene (**35**), β-pinene (**36**), and limonene (**37**), while the components responsible for the stickiness are more complex terpenoids, not found elsewhere in nature, called trinervitenes (for example, **38**) and kempanes (for example, **39**). Nasute soldiers are essentially mobile artillery units. Programmed to confront trouble, they are downright suicidal when individually

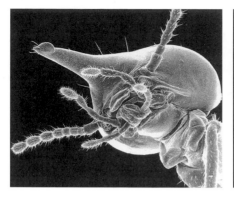

Nasutitermes exitiosus: Left: Head of a soldier. Note the droplet of secretion at the tip of the rostrum. Right: Head of a worker.

challenged, and appear designed for defensive administration of secretion and little else. If individually provoked, a nasute soldier ejects its glandular fluid. The viscous material is squirted out as a fine filament, to a distance of several times the length of the termite. The discharges are aimed, and owing to the revolvability of the termite's head, can be ejected forward, to the sides, and even posteriorly. Soldiers respond only to direct assault. They do not discharge preemptively in response to trouble nearby.

When discharging, the soldiers typically jerk the body front to back, while at the same time moving the head from side to side, with the result that the secretory filament is ejected with loops and inflections over a broad area. The discharges last only 0.1 second or less. After an ejection, the soldier wipes its rostrum against the substrate. A single discharge is usually all it is able to offer.

The secretion is highly effective in its action against ants, spiders, centipedes, and other arthropods. It is on the one hand potently irritating, on account of its monoterpenes, and on the other immobilizing, by virtue of its sticky components, the trinervitenes and kempanes. The combined action of the two sets of components usually suffices to doom arthropod enemies.

The nasute spray also serves as an alarm pheromone. As soon as an enemy is sprayed by a soldier, other soldiers rally to the site and encircle the offender. The new arrivals remain in attendance, and if

35. α-Pinene   **36.** β-Pinene   **37.** Limonene

38. Trinervitene   **39.** Kempane

directly challenged will themselves add their secretion to the target. The discharges eventually take their toll, but it may be some time before an enemy is completely immobilized. The surrounding soldiers hold their post for as long as an enemy shows signs of life. The slight air motions engendered by a twitching animal before it is actually dead are all that is needed to keep the soldiers on the alert.

A simple experiment with an animated mechanical dummy provided clear evidence of how nasute termites coordinate their actions when alerted to an emergency. *N. exitiosus* can be induced to lay a chemical trail in a petri dish, simply by releasing several dozen of the termites into the dish and giving them time. Within an hour or so they will have organized themselves into a column of marchers, running endlessly in a circle just inside the periphery of the dish. When thus occupied, the termites can be subjected to controlled assault. The attacker can take the form of a small metallic bar, made to twirl under the action of a magnet placed beneath the dish. Sliding the dish over the magnet will move the twirler toward the flank of the

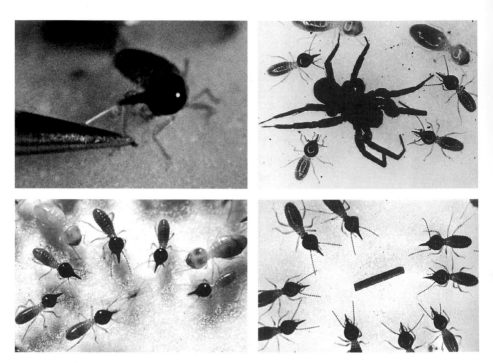

Nasutitermes exitiosus soldiers and workers. Top left: A soldier discharging a jet of secretion in response to the pinching of one of its legs. Top right: A wolf spider, already incapacitated by the spray of soldiers, being guarded by a new set of soldiers ready to add their own spray should the spider show signs of renewed activity. Bottom left: Soldiers and workers, taking action against a rotating metal twirler that has "attacked" them. Bottom right: The twirler has become "incapacitated" by having been glued down by the soldier's spray.

marching termites. As soon as the twirler contacts a soldier, that soldier confronts the device, aims its rostrum, and discharges. Wetted by secretion, the twirler can now be made to attack additional soldiers, which, if themselves hit by the twirler, add their spray to the device. By then the column is totally disrupted, but it usually re-forms along a path that circumvents the site of action. The twirler, for as long as it keeps twirling, remains surrounded by attentive soldiers. Eventually, however, as a consequence of its sticky coating, the twirler becomes fastened to the floor of the dish and ceases twirling, whereupon the soldiers near it lose interest and join the marching column. At this point worker termites may themselves become interested in the "dead" enemy, which they will bury with soil particles if such material is made available to them. Workers may also join in the initial effort to incapacitate the enemy. They have powerful mandibles, which they readily put to use if they come across an assailant. But the bulk of the defensive effort is carried out by the soldiers, which use their secretion both to incapacitate the enemy and to enlist the help of additional soldiers.

The alarming effect of the spray can be demonstrated simply by presenting the termites with isolated samples of the secretion. Secretion can be squeezed readily from the heads of freshly killed nasute soldiers. If a dab of this material is placed next to a column of *N. exitiosus,* or into a chamber of a laboratory colony of the termite, soldiers will promptly converge upon the sample and deploy themselves around it in readiness for action.

In Australia, termites would have little chance of surviving without effective defenses because ants are a dominant component of the Australian fauna and a threat to termites in virtually every habitat. *N. exitiosus* doubtless owes its survival in no small measure to the effective orchestration of its defenses.

Termites are believed to have arisen from a group of primitive wood-eating cockroaches. The order to which they belong, the Isoptera, includes some 2,000 species worldwide, mostly of tropical distribution. Only a few species penetrate far into the temperate zone. The Termitidae, with about 1,500 species, constitute the largest and most diverse family within the order.

REFERENCES

Beaton, C., T. Eisner, and I. Kriston. 1973. Insect behavior: surrogate insects produced with magnetic stirrer. *Annals of the Entomological Society of America* 66:1365–1366.
DeLigne, J., A. Quennedey, and M. S. Blum. 1982. The enemies and defense mechanism of termites. In H. R. Hermann, ed., *Social Insects.* New York: Academic Press.

Eisner, T., I. Kriston, and D. J. Aneshansley. 1976. Defensive behavior of a termite (*Nasutitermes exitiosus*). *Behavioral Ecology and Sociobiology* 1:83–125.

Karlson, P., and M. Lüscher. 1956. "Pheromones": a new term for a class of biologically active substances. *Nature* 183:55.

Moore, B. P. 1968. Studies on the chemical composition and function of the cephalic gland secretion in Australian termites. *Journal of Insect Physiology* 14:33–39.

Noirot, C. H. 1969. Glands and secretion. In K. Krishna and F. M. Weesner, eds., *Biology of Termites*. New York: Academic Press.

Prestwich, G. D. 1974. Chemical defense in termite soldiers. *Journal of Chemical Ecology* 10:1799–1807.

——— 1984. Defense mechanisms of termites. *Annual Review of Entomology* 29:201–232.

19

Class INSECTA
Order PHASMATODEA
Family Diapheromeridae
Oreophoetes peruana
A walkingstick

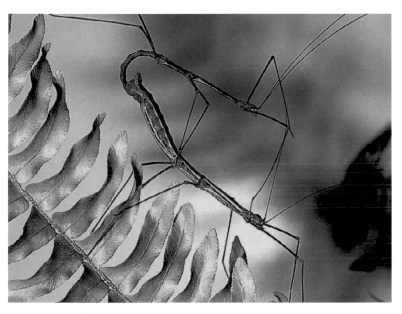

A mating pair of *Oreophoetes peruana.*

Insects of the order Phasmatodea, called phasmatids, are also known as stick insects or walkingsticks. The terms are appropriate because most of the approximately 2,500 species do indeed have an uncanny resemblance to twigs or branches. As entomologists know all too well, and predators undoubtedly come to learn, stick insects can be extraordinarily difficult to spot.

An unidentified stick insect, showing the camouflage that is typical of most Phasmatodea.

The Peruvian walkingstick, *Oreophoetes peruana,* is exceptional in that it is gaudily colored, literally flaunting itself. Since it is flightless and slow on its feet, like stick insects generally, we would naturally suspect that the animal is somehow protected. And indeed it is. *O. peruana* has two sac-like glands in the thorax, present in both sexes and all stages, with openings just behind the head, from which it discharges a malodorous white fluid, even when gently handled. The secretion is a fine aqueous emulsion. The suspended phase consists of a single compound, quinoline (**40**), not known to be produced for defensive purposes by any other animal. It is unusual for the defensive secretion of an arthropod to contain only a single active component. As a rule, defensive secretions are mixtures, containing anywhere from a few to a great many constituents.

Left: Defensive glands (visible as purple sacs) of *Oreophoetes peruana*, exposed by dissection of a dead specimen. Right: Residue of secretion on the prothorax of an *O. peruana* that has just discharged.

Quinoline has the combined attributes of a repellent and an irritant. Ants that have been lured to feed at a sugar solution can be repelled instantly, simply by holding a capillary tube bearing quinoline close to them. Wolf spiders, too, are sensitive. If quinoline is applied to a wolf spider's mouthparts while it is feeding on an insect, the spider ceases feeding and drops the food item. Proof of the irritancy of quinoline can be obtained by applying minute droplets of the chemical to the integument of animals. Frogs respond to such stimulation by promptly scratching the affected site, as do cockroaches. The latter are not natural predators of stick insects, but their topical sensitivity to quinoline could be indicative of a general surface sensitivity of insects to the compound. The question of how it is that *O. peruana* can itself tolerate quinoline remains unanswered.

Although quinoline has not been isolated from other animals, a closely related compound, indole (**41**), is known to be produced, apparently also for defensive purposes, by certain ants, caddisflies, and scarab beetles. Another related compound, naphthalene (**42**), a

40. Quinoline   41. Indole   42. Naphthalene

major constituent of coal tar, is widely used as an insect repellent, not by insects, but by humans. Naphthalene is the active principle in moth balls.

REFERENCES

Bedford, G. O. 1978. Biology and ecology of the Phasmatodea. *Annual Review of Entomology* 23:125–149.

Eisner, T., R. C. Morgan, A. B. Attygalle, S. R. Smedley, K. B. Herath, and J. Meinwald. 1997. Defensive production of quinoline by a phasmid insect *(Oreophoetes peruana)*. *Journal of Experimental Biology* 200:2493–2500.

Floyd, D. 1993. *Oreophoetes peruana*—a very unconventional stick insect. *Bulletin Amateur Entomological Society* 52:121–124.

20

Class INSECTA
Order PHASMATODEA
Family Pseudophasmatidae
Anisomorpha buprestoides
The two-striped walkingstick

Anisomorpha buprestoides: a male astride a female.

The defensive glands of *Anisomorpha buprestoides*, exposed by dissection of a dead insect.

A*nisomorpha buprestoides* is a robust walkingstick from the southeastern United States. The pioneer American entomologist Thomas Say, quoting an acquaintance, reports that this insect, when captured, responds by "diffusing a strong odor." This is an understatement. *A. buprestoides* is the source of one of the most noxious defensive secretions known to be produced by an insect.

A. buprestoides has two large compressible glands in the thorax (similar to but larger than those of *Oreophoetes peruana;* see Chapter 19), from which it expels a fine mist to distances of up to 30 centimeters when disturbed. The gland openings are on the thorax immediately behind the head, and the animal is able to discharge from one gland or both, in virtually any direction. The secretion is potently irritating, and can induce excruciating pain, particularly if it comes in contact with the eyes, but causes considerable discomfort even when merely inhaled.

Like *O. peruana*, *A. buprestoides* discharges an emulsion containing a single active component. The compound, named anisomorphal (or dolichodial) (**43**), is an isoprenoid (technically, a cyclopentanoid monoterpene) potently repellent to insectan and vertebrate predators. The glands are present in both sexes and in all developmental stages, and are even replete with secretion and functional when the insect emerges from the egg.

Anisomorphal is produced also by a mint plant, in which the compound is sealed within tiny capsules embedded in the leaf tissue. The capsules are designed to rupture and to release their repellent contents when herbivores bite into the leaves. A compound closely related to anisomorphal is nepetalactone (**44**), or catnip, well known for its stimulant effect on cats. Whatever the significance of its action on cats, nepetalactone is powerfully repellent to insects and probably active in that capacity in the mint plant that produces it. Interestingly, nepetalactone is also found in the defensive glands of *Graeffea crovani*, another walkingstick.

A. buprestoides is unusual in that it does not necessarily wait until it is directly assaulted before it discharges its spray. When approached by a bird, it often sprays preemptively, before it is actually pecked. It holds its "fire" until the bird is within a radius of about 20 centimeters or less, well within the range of its spray, and then discharges with precise aim. The event is memorable to a bird such as a

43. Anisomorphal   44. Nepetalactone

blue jay, which may learn to shun *A. buprestoides* on the basis of a single experience. Interestingly, *A. buprestoides* does not discharge simply in response to nearby movement. To cause the insect to spray, a stimulus very closely imitative of a bird has to be presented. Waving a colored cloth or a bundle of feathers in the insect's vicinity will not do the trick. The insect is obviously programmed not to waste its secretion by being unduly "trigger happy."

Despite its potent chemical defense, *A. buprestoides* does occasionally fall victim to predation. Examination of the fecal remains of black bears *(Ursus americanus)* in Florida showed remnants of *A. buprestoides,* as well as of another chemically protected insect, the formic acid–spraying ant *Camponotus floridanus* (see Chapter 68).

REFERENCES

Carlberg, U. 1985. Chemical defense in *Anisomorpha buprestoides* (Insecta, Phasmida). *Zoologischer Anzeiger* 215:177–188.

Chow, Y. S., and Y. M. Lin. 1986. Actinidine, a defensive secretion of stick insect, *Megacrania alpheus* Westwood (Orthoptera: Phasmatidae). *Journal of Entomological Science* 21:97–101.

Eisner, T. 1964. Catnip: its raison d'être. *Science* 146:1318–1320.

———1965. Defensive spray of a phasmid insect. *Science* 148:966–968.

Eisner, T., M. Eisner, D. J. Aneshansley, C.-L. Wu, and J. Meinwald. 2000. Chemical defense of the mint plant, *Teucrium marum* (Labiatae). *Chemoecology* 10:211–216.

Meinwald, J., M. S. Chadha, J. J. Hurst, and T. Eisner. 1962. Defense mechanisms of arthropods. IX. Anisomorphal, the secretion of a phasmid insect. *Tetrahedron Letters* 1:29–33.

Meinwald, J., G. M. Happ, J. Labows, and T. Eisner. 1966. Cyclopentanoid terpene biosynthesis in a phasmid insect and in catmint. *Science* 151:79–80.

Paysee, E. A., S. Holder, and D. K. Coats. 2001. Ocular injury from the venom of the Southern walking stick. *Ophthalmology* 108:190–191.

Roof, J. C. 1997. Black bear food habits in the lower Wekiva river basin of central Florida. *Florida Field Naturalist* 25:92–97.

Smith, R. M., J. J. Brophy, J. W. K. Cavill, and N. W. Davies. 1979. Iridodial and nepetalactone in the defensive secretion of the coconut stick insect *Graeffea crovani*. *Journal of Chemical Ecology* 5:727–735.

21

Class INSECTA
Order ORTHOPTERA
Family Romaleidae
Romalea guttata
The eastern lubber grasshopper

Romalea guttata regurgitating.

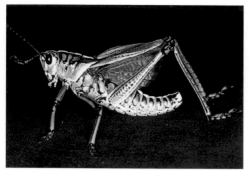

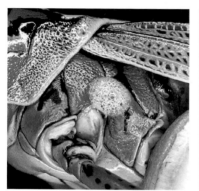

Romalea guttata. Top left: Wing display. Top right: Defensive use of tibial spines. Bottom left: Emission of defensive froth from the left metathoracic spiracle. Bottom center: A toad *(Bufo americanus),* having almost dispatched an adult *R. guttata,* lingers with the grasshopper's legs projecting from the mouth. Bottom right: The head and crop pulled from a *R. guttata* nymph by a captive blue jay *(Cyanocitta cristata)* that went on to eat some of the remainder of the insect.

Like lubber grasshoppers generally, *Romalea guttata* is large and somewhat ungainly. It gives the impression, often conveyed by well-defended animals, that it has "nothing to be concerned about." Conspicuously colored, and slow on its feet, it makes no effort to seek concealment, and although it has wings, these are too short to fulfill their basic function. Lubber grasshoppers cannot fly.

R. guttata exemplifies beautifully what is often true for protected animals: the possession of multiple defenses. *R. guttata* has recourse to four separate protective adaptations. One of these is seldom in evidence, because the insect appears to reserve it for birds. When pecked or merely approached by a bird, *R. guttata* typically flips up its forewings while at the same time unfolding and erecting its ruby-red hindwings. The response, intended to startle and warn, can be elicited by subjecting the grasshopper to a series of quick pinchings with the fingers, just as a bird might do with its bill. Mere live action in the vicinity of the grasshopper does not usually induce the behavior. It takes a realistic imitation of pecking to trigger the action.

The warning implicit in the wing display of *R. guttata* is genuine. Should a predator press its attack upon the grasshopper, it is likely to find itself having to deal with both the sharp spines on the insect's hindlegs and the animal's chemical defenses. The spines are multiple and sharp, and the grasshopper puts them to use by flailing its hindlegs or pushing these against the enemy when attacked. The chemical defenses are twofold. One involves regurgitation of gut contents, and the other consists of emission of a bubbling froth from the metathoracic spiracles. Vomiting is a common defensive response of grasshoppers, and in *R. guttata* it involves emission of fluid in copious amounts. The fluid is noxious and may owe some of its deterrency to substances derived by the insects from their food-plants, which include species known to contain distasteful factors (such as *Eupatorium capillifolium,* which is laden with bitter pyrrolizidine alkaloids). The thoracic froth is emitted with a hiss and, as a consequence of the bursting of its bubbles, results in the formation of a fine, malodorous mist. The froth is the product of secretory tissue associated with the tracheal tubes that lead inward from the metathoracic spiracles. On emission, the liquid produced by this tissue is mixed with tracheal air, hence the frothing. A number of repellent chemicals have been identified in the froth, including phenol (**45**), verbenone (**46**), 1,4-benzoquinone (**30**), isophorone (**47**), and *p*-cresol (**48**).

Other lubber grasshoppers are also defended. *Taeniopoda eques* emits regurgitant and thoracic froth, and manifests a wing display comparable to that of *R. guttata. Brachystola magna* produces massive amounts of oral effluent, but emits no froth and has no wing display.

Despite their defenses, lubber grasshoppers are vulnerable to oc-

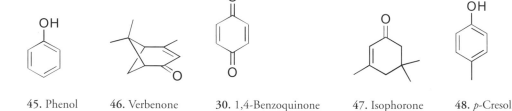

45. Phenol   46. Verbenone   30. 1,4-Benzoquinone   47. Isophorone   48. *p*-Cresol

casional attack. There is evidence, both from laboratory experiments and field observation, that some birds avoid exposure to the noxious enteric contents of lubber grasshoppers by pulling the head and crop from the insect before consuming parts of the remainder of the animal. *R. guttata* also fall victim to loggerhead shrikes, which impale the insects on plant thorns, as they are known to do with other toxic prey as well. They allow the grasshoppers to "age" in this fashion for 1 to 2 days, and then consume them. It is assumed that some of the toxic substances in the grasshopper dissipate during the aging period, or undergo degradation, with the result that the insect is then rendered palatable.

Lubber grasshoppers are members of the orthopteran family Romaleidae. The family includes about 80 genera and 200 described species. Most are found in Central and South America, but a few occur in North America.

REFERENCES

Carpenter, G. D. H. 1938. Audible emission of defensive froths by insects. *Proceedings of the Zoological Society, London* 108:243–252.
Eisner, T. 1970. Chemical defense against predation in arthropods. In E. Sondheimer and J. B. Simeone, eds., *Chemical Ecology.* New York: Academic Press.
Freeman, M. A. 1968. Pharmacological properties of the regurgitated crop fluid of the African migratory locust, *Locusta migratoria* L. *Comparative Biochemistry and Physiology* 202:1041–1049.
Hatle, J. D. 1998. Slow movement increases the survivorship of a chemi-

cally defended grasshopper in predatory encounters. *Oecologia* 115:260–267.

Jones, C. G., T. A. Hess, D. W. Whitman, P. J. Silk, and M. S. Blum. 1987. Effects of diet breadth on autogenous chemical defense of a generalist grasshopper. *Journal of Chemical Ecology* 13:283–297.

Lymbery, A., and W. Bailey. 1980. Regurgitation as a possible anti-predator defense mechanism in the grasshopper *Goniaea* sp. (Acrididae: Orthoptera). *Journal of the Australian Entomological Society* 19:129–130.

Sword, G. A. 2001. Tasty on the outside, but toxic in the middle: grasshopper regurgitation and host plant–mediated toxicity to a vertebrate predator. *Oecologia* 128:416–421.

Whitman, D. W., and L. J. Orsak. 1985. Biology of *Taeniopoda eques* (Orthoptera Acrididae) in Southeastern Arizona, USA. *Annals of the Entomological Society of America* 78:811–825.

Whitman, D. W., M. S. Blum, and D. W. Alsop. 1990. Allomones: chemicals for defense. In D. L. Evans and J. O. Schmidt, eds., *Insect Defenses*. Albany: State University of New York Press.

Yosef, R., and D. W. Whitman. 1992. Predator exaptation and defensive adaptation in evolutionary balance: no defense is perfect. *Evolutionary Ecology* 6:527–536.

22

Class INSECTA

Order HEMIPTERA

Family Coreidae

Chelinidea vittiger

A leaf-footed bug

A freshly molted *Chelinidea vittiger,* above its shed cuticle, on a spine of its *Opuntia* foodplant.

Chelinidea vittiger. Top row: A nymph emerging from the egg. Bottom left: A half-grown nymph, showing the gland openings on the back of the abdomen. Bottom right: An adult, showing the right gland opening on the flank of the thorax.

The Hemiptera, one of the most diverse of insect orders, comprise some 50,000 species placed in about 100 families. The mouthparts of hemipterans characteristically are formed into a rostrum or proboscis, specialized for fluid uptake. A number of hemipteran families, notably the Coreidae, or leaf-footed bugs, and the Pentatomidae, or stink bugs, have defensive glands from which they eject strongly odorous fluids when disturbed. The glands responsible for production of these odorants are similar in coreids and pentatomids. The defenses occur in both sexes of these bugs, and in the immatures as well as the adults. The coreid *Chelinidea vittiger* provides a typical example.

C. vittiger occurs commonly on prickly pear cactus (genus *Opuntia*) in central Florida. In the adult bug, there is a single defensive gland, which takes the form of a large sac, positioned mid-ventrally in the thorax and opening by way of two pores positioned on either side, just above the base of the third leg. Emission through the pores

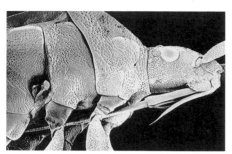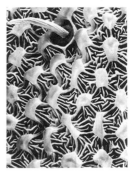

Chelinidea vittiger. Left: An adult in lateral view, showing the right gland opening. Middle: An enlarged view of the gland opening (note the stippled appearance of the cuticule in the surrounds of the orifice). Right: Detail of the cuticule adjacent to the gland opening, showing the microsculpting that accounts for the cuticle's stippled appearance at a lower magnification.

is controlled by special valves. The animal is thus capable of discharging from one pore or the other, or from both simultaneously. The secretion is ejected in the form of coarse droplets, rather than as a fine spray.

The cuticle in the immediate surrounds of each glandular pore is minutely sculpted. The roughness ensures that the region becomes soaked in secretion following each discharge. The bug is thus given extended protection. Such sculpted zones are a standard feature of Coreidae and Pentatomidae.

The secretion in an adult *C. vittiger* contains a mixture of repellent compounds, including *(E)*-2-hexenal (**28**), acetic acid (**1**), and *n*-hexyl acetate (**49**).

Immature *C. vittiger* (the nymphs) have dorsal abdominal defensive glands, in lieu of the thoracic gland present in the adult. These nymphal glands open on the back of the abdomen between segments 4 and 5, and 5 and 6. The difference in glandular arrangement between nymph and adult makes sense, because retaining the nymphal glands in the adult would mean having the gland openings beneath the wings. Creating a new arrangement in the adult, whereby discharges can be effected from the flanks of the thorax, solves the problem.

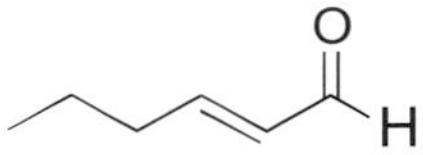

28. *(E)*-2-Hexenal

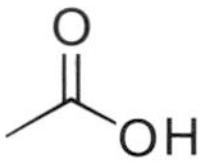

1. Acetic acid

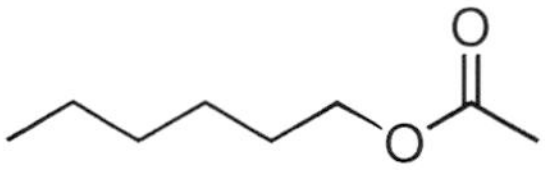

49. *n*-Hexyl acetate

50. Tridecane

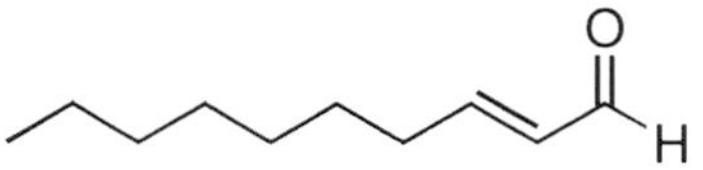

51. *(E)*-2-Decenal

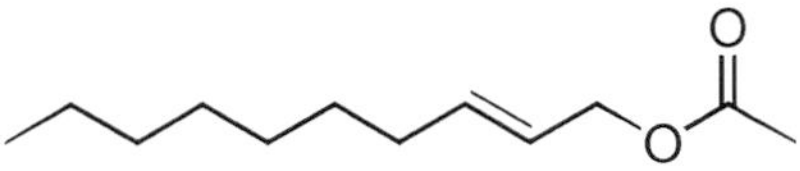

52. *(E)*-2-Decenyl acetate

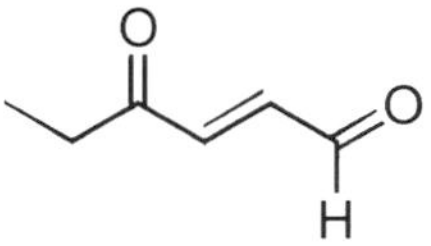

53. *(E)*-4-Oxo-2-hexenal

In pentatomids there is also, characteristically, a single thoracic gland in the adult, and multiple dorsal abdominal glands in the nymphal stages. Some very thorough work has been done on the secretion of pentatomids. In *Nezara viridula,* a species stemming from the Ethiopian region that has spread worldwide and become a pest of crops in many regions, the defensive secretion in the adult contains a mixture of compounds, including the hydrocarbon tridecane **(50)**, the aldehydes *(E)*-2-hexenal **(28)** and *(E)*-2-decenal **(51)**, the ester *(E)*-2-decenyl acetate **(52)**, and the keto aldehyde *(E)*-4-oxo-2-hexenal **(53)**. The secretion is effectively repellent to some predators, but is apparently only moderately effective against birds. Some parasites, too, appear to be tolerant of the secretion. One parasitoid, the tachinid fly *Trichopoda pennipes,* seems even to be attracted chemically to *N. viridula,* not to components of the defensive secretion, but to certain pheromonal constituents that mediate sex attrac-

A *Nezara viridula* nymph (left) and adult (right), on a legume plant *(Crotalaria mucronata).*

tion in the bug. Another parasitoid, the scelionid wasp *Trissolcus basalis,* which attacks the eggs of *N. viridula,* appears to locate the bugs by being attracted to the defensive products of the adult.

The abdominal defensive glands in the nymphs of *N. viridula* produce a mixture differing from that of the adult, but based on the same general categories of compounds.

Other Hemiptera, including, for example, members of the families Cydnidae and Scutelleridae, have glandular systems comparable to those of the Coreidae and Pentatomidae.

Coreids and pentatomids are successful groups, in part, no doubt, because of their effective defenses. The Coreidae contain some 250 genera with about 1,800 species; the Pentatomidae include about 400 genera with 5,000 species.

REFERENCES

Aldrich, J. R. 1988. Chemical ecology of the Heteroptera. *Annual Review of Entomology* 33:211–238.

———— 1995. Chemical communication in the true bugs and parasitoid exploitation. In R. T. Cardé and W. J. Bell, eds., *Chemical Ecology of Insects, II.* New York: Chapman and Hall.

DeVol, J. E., and R. D. Goeden. 1973. Biology of *Chelinidea vittiger* with notes on its host-plant relationship and value in biological weed control. *Entomologist* 2:231–240.

Filshie, B. K., and D. F. Waterhouse. 1968. The fine structure of the lateral scent glands of the green vegetable bug *Nezara viridula* (Hemiptera, Pentatomidae). *Journal de Microscopie* 7:231–244.

———— 1968. The fine structure of a surface pattern of the cuticle of the green vegetable bug *Nezara viridula. Tissue and Cell* 1:367–385.

Gilby, A. R., and D. F. Waterhouse. 1965. The composition of the scent of the green vegetable bug, *Nezara viridula. Proceedings of the Royal Society* B 162:105–120.

Mattiacci, L., S. B. Vinson, H. J. Williams, J. R. Aldrich, and F. Bin. 1993. A long-range attractant kairomone for egg parasitoid *Trissolcus basalis,* isolated from defensive secretion of its host, *Nezara viridula. Journal of Chemical Ecology* 6:1167–1181.

McCullogh, T. 1974. Chemical analysis of the defensive scent fluid of the cactus bug, *Chelinidea vittiger. Annals of the Entomological Society of America* 67:300.

Remold, H. 1962. Über die biologische Bedeutung der Duftdrüsen bei den Landwanzen (Geocorisae). *Zeitschrift für Vergleichende Physiologie* 45:636–694.

23

Class INSECTA

Order HEMIPTERA

Family Reduviidae

Apiomerus flaviventris

A reduviid bug

Apiomerus flaviventris feeding on a click beetle.

Members of the family Reduviidae include some highly successful predators that feed largely on insects. They constitute over 5,000 species and about 1,000 genera. They are worldwide in distribution, but occur primarily in the tropics and subtropics.

Reduviids kill their prey by injecting venomous salivary fluid with the beak. They also may use the beak in defense, and reduviid bites can be extremely painful. One African species, *Platymeris rhadamanthus,* is exceptional in that it uses its beak as a spray gun. It ejects its saliva in a series of short, aimed squirts, to distances of several feet, thereby holding vertebrate predators at bay. When disturbed, reduviids also commonly emit strong odors, arising from secretions they discharge in small quantity from special glands. The fluids may be directly deterrent to predators, but they could serve also, by virtue of their odors, as alerting signals, to warn enemies of the reduviid's biting ability.

Apiomerus flaviventris possesses all of these defensive attributes, but in addition has an elaborate behavior involving the procurement of plant resin for protection of its eggs. The adult female visits a number of plants, including, in southern Arizona, the common roadside weed *Heterotheca psammophila* (family Asteraceae). The leaves, stems, and buds of this plant are covered with tiny glandular hairs bearing droplets of a sticky, highly aromatic secretion at the tip. This resin is the plant's protection and it is effective: few insects feed on *H. psammophila. A. flaviventris* is exceptional in that it is undeterred by the resin. The adult female visits the plant and harvests the droplets. By scraping with the forelegs, she gathers dabs of the resin, which she then transfers to the midlegs, and from the midlegs, by use of the hindlegs, to the ventral surface of the abdomen. The female may invest hours in this activity, procuring substantial amounts of the resin. Subsequently, when she lays her eggs, she coats each one with the sticky material, and by using her hindlegs glues it to those already laid. The result is a cluster of eggs, all neatly fastened to one another, and protected collectively by a thick coating of resin on the outside of the cluster. The resin is a complex mixture of terpenes, deterrent to ants and other insects, and possibly inhibitory also to microbial pathogens. *A. flaviventris* provides but one example of secondary utilization by an insect, for protective purposes of its own, of the chemical weaponry of a plant. Such defensive opportunism is of frequent occurrence in insects, although the strategies may differ greatly (see Chapters 25, 62, 66, 67).

Left: *Heterotheca psammophila.* Top right: A leaf of *H. psammophila,* showing the secretory droplets produced by glandular hairs. Bottom right: An enlarged view of two glandular hairs.

When *A. flaviventris* nymphs emerge from the eggs, they promptly begin gathering resin from the outer surface of the egg cluster. They use their legs for the purpose, and eventually succeed in coating the tibiae of the forelegs with large dabs of resin. The acquired resin may protect the young against predation, but it appears also to be used by them as an aid in prey capture. The newly hatched *A. flaviventris* are predators like the adults, and they capture prey by grasping it with the forelegs. The resin seems to enable them to hold on to relatively large prey, which might slip from their grasp unless firmly restrained.

Top: An *Apiomeris flaviventris* female, scraping secretion from *Heterotheca psammophila* with its forelegs (left), and subsequently (top right) applying the material to the ventral surface of the abdomen with the hindlegs. Bottom left: A freshly laid egg cluster of *A. flaviventris* coated with resin. Bottom right: A newly emerged *A. flaviventris,* bearing *H. psammophila* secretion on its forelegs that it scraped from the egg cluster.

REFERENCES

Edwards, J. S. 1961. The action and composition of the saliva of an assassin bug *Platymeris rhadamanthus* Gaerst (Hemiptera: Reduviidae). *Journal of Experimental Biology* 38:61–77.

———— 1962. Spitting as a defensive mechanism in a predatory reduviid. *Proceedings of the XI International Congress of Entomology* 3:259–263.

Eisner, T. 1988. Insekten als fürsorgliche Eltern. *Verhandlungen der Deutschen Zoologischen Gesellschaft* 81:9–17.

Eisner, T., C. Rossini, A. González, V. K. Iyengar, M. V. S. Siegler, and S. R. Smedley. 2002. Paternal investment in egg defense. In M. Hilker and T. Meiners, eds., *Chemoecology of Insect Eggs and Egg Deposition*. Berlin: Blackwell Publishing.

24

Class INSECTA
Order HEMIPTERA
Family Belostomatidae
Abedus herberti
A giant water bug

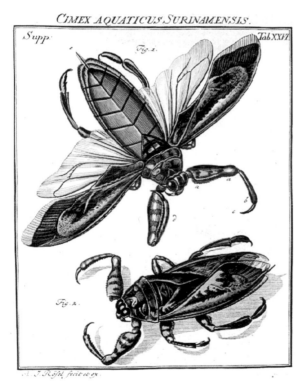

A belostomatid bug with wings folded, as at rest, and with wings spread, as in flight. Illustration by Rösel von Rosenhof, one of the great entomological artists of the eighteenth century. From *Insecten-belustigung,* vol. 3, fascicle 26 (Nuremberg: J. J. Fleischmann, 1755).

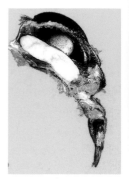

The giant water bug *Lethocerus americanus*. Left: Frontal view of the head, show-ing white secretion (arrow) emanating from the rostral gland. Middle: The head in lateral view. Right: The head dissected open, to show the rostral gland, replete with white secretion.

The Belostomatidae, or giant water bugs, include a number of spe-cies of formidable appearance and nasty reputation. Aquatic hunt-ers, belostomatids capture prey with their raptorial forelegs and quickly immobilize it with poison that they inject with their beak. Belostomatids do not hesitate to use the beak in defense, as many a curiosity-driven human has learned who, upon encountering one of these bugs in an outdoor swimming pool, was reckless enough to handle it. The bite of a belostomatid is very painful. The poison is well suited for incapacitation of prey. Belostomatids routinely feed on insects, fish, and tadpoles, including individuals considerably larger than themselves.

Some belostomatids, including the giant species of the genus *Lethocerus* and the somewhat smaller ones of the genus *Abedus,* have a pair of glands opening at the base of the beak, from which they discharge a white fluid when manually disturbed. The fluid is un-doubtedly defensive, although there is no direct proof of this. Ac-tual predation tests with belostomatids appear not to have been done. But in one species, *Abedus herberti,* a belostomatid from the southwestern United States, the secretion has been studied chemi-cally and found to contain four steroids: desoxycorticosterone (54), pregnenolone (55), progesterone (56), and 3-epipregnenolone (57). All four compounds belong to a category of steroids called preg-

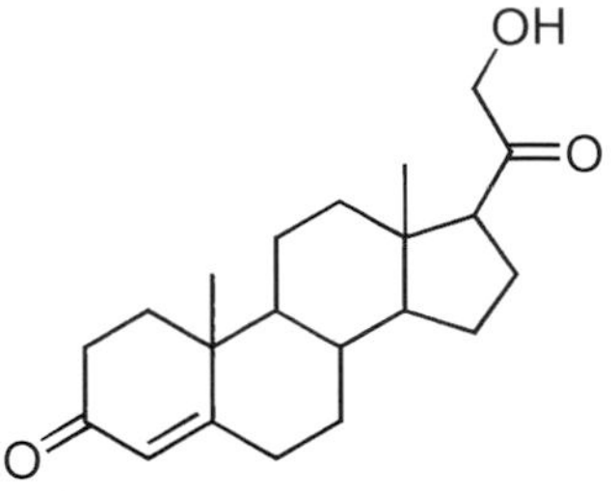

54. Desoxycorticosterone

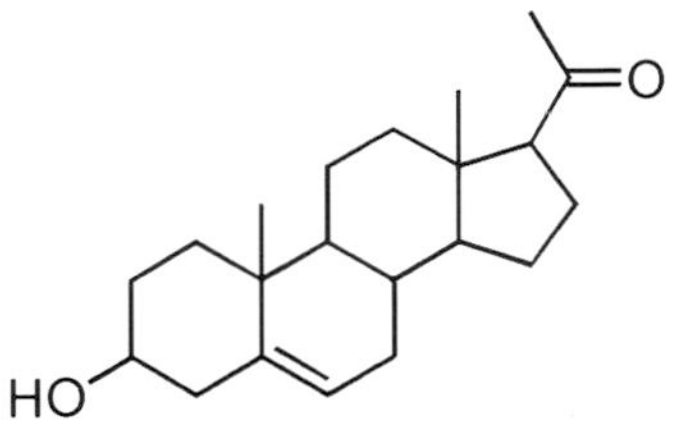

55. Pregnenolone

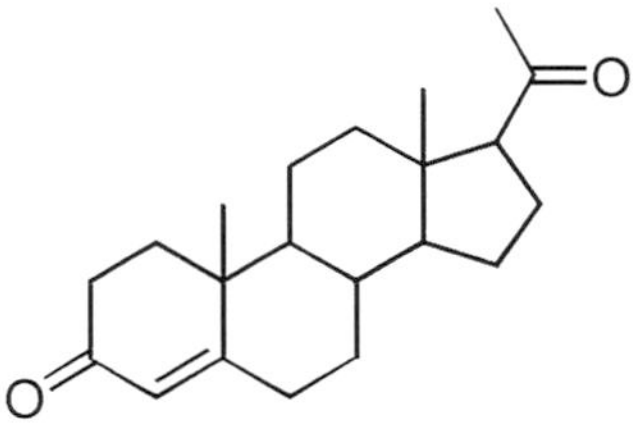

56. Progesterone

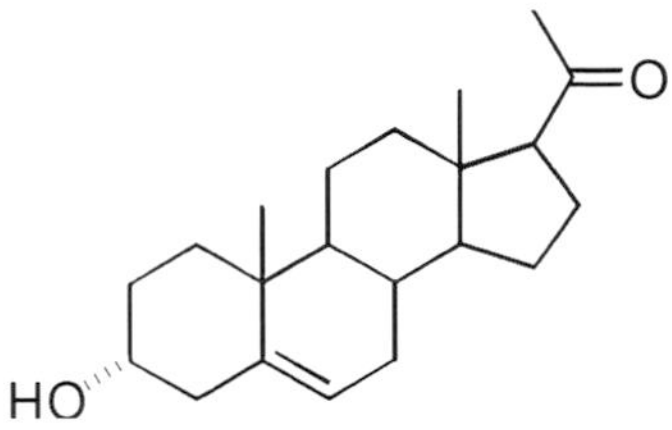

57. 3-Epipregnenolone

nanes, which had previously been characterized from the defensive secretions of other insects, notably acquatic beetles of the family Dytiscidae (see Chapter 37). Both *A. herberti* and dytiscids produce their pregnanes in high concentration, and one component, desoxy-corticosterone, the principal pregnane in the *A. herberti* secretion, had itself been isolated from the secretion of some dytiscids. Des-oxycorticosterone has been tested in feeding experiments with fish, and found to be potently deterrent to these animals. It seems very possible therefore that the pregnanes of belostomatids are protective. In fact, the production by both dytiscids and belostomatids of the same type of defensive steroids, in an aquatic setting where fish con-stitute a major common threat, provides a striking example of paral-lel evolution.

Pregnanes have important hormonal roles in vertebrates, and some of the compounds in the *A. herberti* secretion (for instance, progesterone) are themselves well-known hormonal entities. It is in-teresting that dytiscids and *A. herberti* both produce their defensive pregnanes at concentrations well above what are considered ordinary hormonal levels. A single *A. herberti,* for example, contains on average as much as 1 milligram of desoxycorticosterone. In such a large quantity, pregnanes can elicit prompt and rather drastic physiological effects. Thus, for instance, high oral dosages of pregnanes, aside from being distasteful, can be emetic or even sedative to fish, with the result that dytiscids are sometimes vomited after ingestion or are simply able to crawl out of the mouth of a fish that has become immobilized. At hormonal concentrations pregnanes would not induce such effects.

The production of steroids by insects raises interesting questions, because insects lack the ability to synthesize steroids, except from other steroids. It is assumed that dytiscids and belostomatids both synthesize their defensive steroids from cholesterol, a substance that these predators can be assumed to acquire in substantial quantity from the animals they eat.

Pregnanes are also secreted by the defensive glands of some terrestrial insects, as for example carrion beetles of the family Silphidae (see Chapter 38).

Belostomatids are a relatively small group. The family contains only 8 genera and approximately 100 species worldwide. In some species there is an interesting defensive behavior, involving the brooding of eggs by the males. Sometimes the eggs are carried by the males on their backs; in other instances the females lay the eggs on vegetation and the males remain in attendance close by.

REFERENCES

Ichikawa, N. 1990. Egg mass destroying behavior of the female giant water bug *Lethocerus deyrollei* Vuillefroy (Heteroptera, Belostomatidae). *Journal of Ethology* 8:5–12.

Lauck, D. R., and A. S. Menke. 1961. The higher classification of the Belostomatidae. *Annals of the Entomological Society of America* 54:644–657.

Lokensgard, J., R. Smith, T. Eisner, and J. Meinwald. 1993. Pregnanes from defensive glands of a belostomatid bug. *Experientia* 49:175–176.

25

Class INSECTA
Order HEMIPTERA
Family Aphididae
Aphis nerii
The oleander aphid

An *Aphis nerii* colony on a milkweed plant.

58. Calotropin

Aphids are among the most numerous of insects and among those most destructive to plants. They are also among the most difficult to control. There are some 3,500 species of aphids worldwide, in over 350 genera. About 1,400 species occur in North America.

Aphis nerii, known as the oleander aphid, feeds on a diversity of plants, not only oleander *(Nerium oleander)* but species of Asclepiadaceae (milkweeds) as well. From both of these botanical sources it derives certain toxic steroids, called cardiac glycosides or cardenolides, exemplified by calotropin (**58**), which it accumulates systemically and which provides it with protection.

Like the majority of aphids, *A. nerii* has a pair of peg-like structures called cornicles that project from the rear of the abdomen. The cornicles typically emit droplets of fluid when the aphids are disturbed. The fluid apparently consists of wax mixed with an aqueous carrier. The wax is liquid when discharged, but it solidifies promptly on contact with an external surface. It has been suggested that the wax is initially in a liquid crystalline state (supercooled) and that it changes to the solid crystal phase on contact with a seeding nucleus (just as supercooled water crystallizes if abruptly stirred or if it is "seeded" by having a piece of ice dropped into it). The melting point of cornicle wax (37.5°–48°C, depending on the species of aphid) is higher than normal summer temperatures, which indicates that crystallization by seeding can occur under ambient conditions.

Cornicle wax provides effective protection against parasitic Hy-

Top left: *Aphis nerii* emitting a droplet of cornicle wax in response to being poked with a metal probe. Top right: A braconid wasp, its abdomen bent forward, about to oviposit in an aphid. Bottom left: A similar wasp, stuck to the plant's surface, its legs and antennae contaminated with cornicle wax. Bottom right: A similar wasp stuck by its legs to solidified cornicle wax.

menoptera. Braconid wasps, for instance, may find themselves partially or totally immobilized by the wax droplets, which the aphids are quick to emit when contacted by the wasps. It is not uncommon to find dead parasitoid wasps, fastened to branches by cornicle wax, in the midst of aphid colonies.

The cornicle secretion in some aphids has been shown to be the source of an alarm pheromone released by disturbed aphids upon emission of their cornicle droplets, a signal that alerts undisturbed individuals to the imminence of danger.

Despite its cornicles and systemic cardenolides, *A. nerii* has multiple enemies that can cope with its defenses. Among these enemies are larval and adult coccinellid beetles and syrphid fly larvae.

Three predators—a coccinellid beetle (left), a coccinellid larva (middle), and a syrphid fly larva (right)—feeding on *Aphis nerii,* undeterred by the aphids' defense.

REFERENCES

Dixon, A. F. G. 1958. The escape response shown by certain aphids to the presence of the coccinellid *Adelia decempunctata* (L.). *Transactions of the Royal Entomological Society, London* 110:319–334.

Edwards, J. S. 1966. Defence by smear: super-cooling in the cornicle wax of aphids. *Nature* 211:73–74.

Hottes, F. C. 1928. Concerning the structure, function, and origin of the cornicles of the family Aphidae. *Proceedings of the Biological Society of Washington* 41:71–84.

Malcolm, S. B. 1989. Disruption of web structure and predatory behavior of a spider by plant-derived chemical defenses of an aposematic aphid. *Journal of Chemical Ecology* 15:1699–1716.

Outreman, Y., A. Le Ralec, M. Plantegenest, B. Chaubet, and J. S. Pierre. 2001. Superparasitism limitation in an aphid parasitoid: cornicle secretion avoidance and host discrimination ability. *Journal of Insect Physiology* 47:339–348.

Rothschild, M. 1973. Secondary plant substances and warning colouration in insects. In H. F. von Emden, ed., *Insect/Plant Relationships.* Oxford: Blackwell Scientific Publications.

Rothschild, M., J. von Euw, and T. Reichstein. 1970. Cardiac glycosides in the oleander aphid. *Journal of Insect Physiology* 16:1141–1145.

Strong, F. E. 1967. Observations on aphid cornicle secretions. *Annals of the Entomological Society of America* 60:668–673.

26

Class INSECTA

Order HEMIPTERA

Family Aphididae

Prociphilus tessellatus

The woolly alder aphid

A colony of *Prociphilus tessellatus* on an alder bush *(Alnus rugosa).*

Among aphids there are some, called woolly aphids, whose bodies bear a dense covering of waxy filaments, extruded as tufts from clusters of integumental secretory cells. The "wool," which imparts a distinctly fluffy appearance to these aphids, is defensive. Insect predators that bite into woolly aphids may succeed in grasping only wax, and when they do they are induced to clean themselves and break off their assault. Parasitoid wasps may likewise be deterred by the

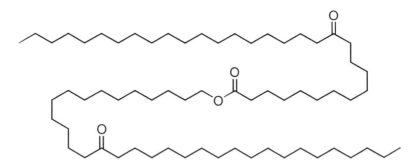

59. 15-Oxotetratriacontanyl 13-oxodotriacontanoate

wax. Woolly aphids lack cornicles, the spigot-like wax-delivery devices from which so many other aphids derive protection (see Chapter 25). The waxy cover of woolly aphids appears to serve as an alternative to the cornicles.

The wool of the aphid *Prociphilus tessellatus,* a species in which the waxy cover has been studied chemically, consists of a single compound, a keto ester, characterized as 15-oxotetratriacontanyl 13-oxodotriacontanoate (**59**). The mechanism by which the wax is extruded from the secretory cells and forced into filaments is not fully understood. The wool of *P. tessellatus* doubtless serves for defense, but the aphid also derives advantages, as do many aphids, from the protective services that it obtains from ants. Aphids pump large quantities of plant sap through their bodies, in order to procure sufficient nitrogenous compounds for sustenance and for production of offspring. This fluid uptake provides them with more carbohydrate than they need, and they void this excess as part of a fluid called honeydew, which they excrete through the anus and offer to the ants in "payment" for their services. The ants eagerly partake of this fluid as it emerges from the aphids' rear, and they are extremely aggressive when thus engaged. You can verify this by poking aphids with your finger. If ants are in attendance, they will promptly turn upon the finger and attempt to attack it. The ants direct their aggression not only against enemies of the aphid but also against competitors intent on imbibing honeydew. Wasps that attempt to partake of the effluent may be driven away by the ants.

Top left: Individuals of *Prociphilus tessellatus,* closely clustered as they typically are in a natural colony. Bottom left: A guarding ant drinking a droplet of honeydew. Right: A guarding ant responding to an "attacking" finger.

The defensive use of ants by aphids provides an example of what is relatively rare in nature: the exploitation by one animal of the defensive capacities of another animal (see Chapters 29, 41, and 48). It is more usual for animals to make use of plant defenses, as exemplified by the many insects that derive protection from ingested plant toxins.

Its defenses notwithstanding, *P. tessellatus* has its share of enemies. Among these are a number of syrphid fly larvae and the caterpillar of a butterfly, *Feniseca tarquinius.* There is also the larva of a green lacewing, which avoids being attacked by the tending ants by cloaking itself with "wool" that it plucks from the aphids, thereby acquiring the appearance of an aphid and being mistaken for an aphid by the ants (see Chapter 33).

REFERENCES

Agarwala, B. K. 2001. Larval interactions in aphidophagous predators: Effectiveness of wax cover as defence shield of *Scymnus* larvae against predation from syrphids. *Entomologia Experimentalis et Applicata* 100:101–107.

Dixon, A. F. G. 1985. *Aphid Ecology.* Glasgow: Blackie & Son.

Dorset, D. L., and H. Ghiradella. 1983. Insect wax secretion: the growth of tubular crystals. *Biochimica et Biophysica Acta* 760:136–142.

Meinwald, J., J. Smolanoff, A. C. Chibnall, and T. Eisner. 1975. Characterization and synthesis of waxes from homopterous insects. *Journal of Chemical Ecology* 1:269–274.

Pike, N., D. Richard, N. Foster, and L. Mahaderan. 2002. How aphids lose their marbles. *Proceedings of the Royal Society London* B 269:1211–1215.

Smith, R. G. 1999. Wax glands, wax production, and the functional significance of wax use in three aphid species (Hemiptera: Aphididae). *Journal of Natural History* 33:513–530.

27

Class INSECTA
Order HEMIPTERA
Family Flatidae
Ormenaria rufifascia
A flatid planthopper

An adult *Ormenaria rufifascia.*

The Flatidae are one of a group of hemipteran families known as planthoppers. And hop they do, both as adults and nymphs. The adults take to the wing by leaping into the air, and the nymphs resort to hopping when changing location or making an escape.

Ormenaria rufifascia is a flatid found on palmetto plants (for example *Sabal palmetto, S. etonia, Serenoa repens*) in the southeastern

Top left: Waxy pads of *Ormenaria rufifascia* nymphs on a palmetto *(Sabal palmetto)*. Top middle and right: Close-up view of a waxy pad and of the resident *O. rufifascia* nymph. Bottom left: An ant that has become contaminated with wax after having been coaxed to traverse an *O. rufifascia* pad. Bottom right: A nymph of *O. rufifascia* infested with *Caeculisoma* mites (note the virtual absence of wax on the bug).

United States. The adult has the wedge-shaped appearance and large triangular forewings typical of flatids. Far from being camouflaged, the animal is quite conspicuously colored, as if it were advertising itself. And well it might be, since with its prodigious leaping ability it is very difficult to catch. Tropical flatids, which are equally prone to jump, tend also to be brilliantly colored.

In leaping, both adults and nymphs of *O. rufifascia* make use of spines on their hindlegs to gain the purchase necessary for takeoff. The spines are on the tibiae and are a characteristic of the Flatidae.

The nymphs of *O. rufifascia* have an additional defense that involves use of the filamentous wax that they secrete from the time they emerge from the egg. The nymphs secrete these filaments as tufts, from glandular tissue in the abdomen. Two of these tufts project directly backward from the tip of the abdomen, the other two protrude obliquely to the sides. The nymphs wipe these tufts against the palmetto leaf surface in their immediate surroundings, thereby creating waxy pads within which they establish residence. Conspicuously white, these pads are easily spotted from a distance.

The *O. rufifascia* nymphs inhabit the pads singly or in groups. When touched they usually hop away, but they may only rarely experience direct assault. The wax surrounding them is an effective barrier, which enemies such as ants and predaceous coccinellid beetle larvae are reluctant to cross. If under experimental conditions an ant is forced to traverse an *O. rufifascia* pad, it becomes topically contaminated with the wax and consequently compelled to engage in prolonged preening activities. Thus distracted, the ant usually desists from pursuing its assault.

Leaping is an effective escape tactic for *O. rufifascia* nymphs. It takes place abruptly and involves the expenditure of considerable muscular energy. Even a mid-size nymph less than 5 millimeters in length can jump distances of 20–30 centimeters. The waxy body coating helps in this regard. If in the midst of a leap a nymph strikes a spider web, it is prevented by its waxy covering from sticking to the viscid strands of the orb (see Chapters 30 and 62). The wax also offers protection should a nymph leap into a puddle. *Ormenaria* nymphs float on water and can use their legs to paddle their way to dryness.

Like many insects, *O. rufifascia* can become infested with mites. One species, a red velvet mite of the genus *Caeculisoma* (family Erythraeidae), can cause an *O. rufifascia* to curtail its wax production and to be less prone to jump when assaulted. The parasitism is evidently costly to the host.

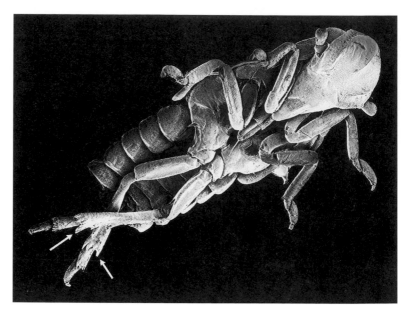

A scanning electronmicrograph of an *Ormenaria rufifascia* nymph leaping. The insect was immobilized in mid-leap by being forced to jump into a refrigerated fluid. Note the spines on the tibiae of the hindlegs (arrows) used by the animal to push itself off for the jump.

There is also a predator, the larva of a green lacewing, *Ceraeochrysa smithi,* that readily feeds on *O. rufifascia.* It is apparently undeterred by the waxy pads.

There are some 1,000 species of Flatidae worldwide, placed in 212 genera.

REFERENCES

LaMunyon, C., and T. Eisner. 1990. Effect of mite infestation on the anti-predator defenses of an insect. *Psyche* 97:31–41.

Wilson, S. W., and J. H. Tsai. 1984. *Ormenaria rufifascia* (Homoptera: Fulgoroidea: Flatidae): descriptions of nymphal instars and notes on field biology. *New York Entomological Society* 82:307–315.

28

Class INSECTA
Order HEMIPTERA
Family Cercopidae
Prosapia bicincta
The two-lined spittlebug

Prosapia bicincta, seized in forceps, responding by emitting droplets of blood from the tarsi (inset: enlarged view of one such tarsus).

The Cercopidae, like the Flatidae (see Chapter 27), leap when disturbed. Known as froghoppers as adults, they tend to respond instantly when threatened by leaping into the air and flying away. Adult cercopids are sometimes also protected chemically. *Prosapia*

Left: Mass of froth, produced by an unidentified cercopid nymph. Middle: The froth has been cleared, exposing the resident nymph. Right: A cercopid nymph, viewed from the rear, in the process of generating froth.

bicincta, for example, when seized emits yellowish droplets (apparently blood) from the tips of its legs. The fluid appears not to have been studied chemically, but is most probably protective.

As nymphs, cercopids are known as spittlebugs, in reference to the masses of "spittle" they produce that almost hide them from view. The spittle consists of a froth generated by the nymph by mixing respiratory air, emitted from anterior spiracles, with fluid released from the anus. The froth is said to shield the nymphs from desiccation, but it could also act as a deterrent to spiders and ants. The fluid contains no irritants or repellents, but it could act physically to slow a predator in its initial probings, forewarning the nymph within and inducing it to leap away. The nymphs are indeed capable of leaping, but they usually refrain from doing so unless forced out of their foamy enclosure.

Cercopid nymphs, hidden within their froth, sometimes look like tiny wads of cotton. When abundant, they may impart a festooned appearance to the landscape.

The family Cercopidae is worldwide in distribution. It contains about 2,300 species, in over 30 genera.

REFERENCES

Mello, M. L. S., E. R. Pimentel, A. T. Yamada, and A. Storopoli-Neto. 1987. Composition and structure of the froth of the spittlebug, *Deois* sp. *Insect Biochemistry* 17:493–502.

Metcalf, Z. P. 1960–1962. *General Catalogue of the Homoptera.* Fascicle VII, *Cercopoidea.* Raleigh: North Carolina State College.

Peck, D. C. 2000. Reflex bleeding in froghoppers (Homoptera: Cercopidae): variation in behavior and taxonomic distribution. *Annals of the Entomological Society of America* 93:1186–1194.

29

Class INSECTA
Order HEMIPTERA
Family Dactylopiidae
Dactylopius confusus
A cochineal bug

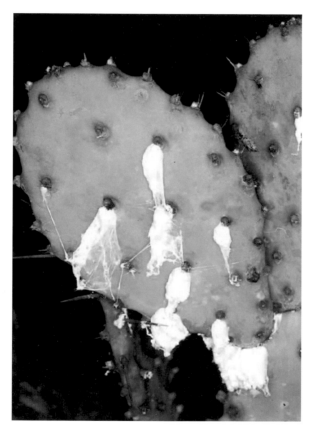

A *Dactylopius confusus* colony on an *Opuntia* cactus.

60. Carminic acid

The insects of the genus *Dactylopius* have an interesting history. Known as cochineal insects, they have been famous as the source of a red pigment, carminic acid or cochineal red (**60**), which for over three centuries, from the mid 1500s to the late 1800s, was the primary red dye used in textile production. The principal species grown for commercial production of the dye was *Dactylopius coccus.* This was raised on its native hosts, prickly pear cacti of the genus *Opuntia,* and harvested by the ton for export purposes in countries such as Mexico. North Africa and the Canary Islands also became involved in cochineal production, and the industry became a mainstay of the economy of these regions. The cochineal industry eventually collapsed in the late 1800s when the dyes began to be produced synthetically. Cochineal red is still used in the cosmetic industry and as a food coloring agent, and *Dactylopius* are still grown in the Canary Islands, but worldwide production of the insect is now minimal in comparison with what it used to be.

Dactylopius confusus, a species occurring in Florida, has a life cycle typical of scale insects in the family Dactylopiidae. The adult females lack legs and wings, and lead a sedentary existence, attached by their mouthparts to the host cactus. The much smaller males do have legs and wings. Dispersal occurs in large measure by way of the newly emerged nymphs, which are ambulatory and active.

Carminic acid is obtained commercially by extracting it from the adult *Dactylopius* females. These are intensely red on account of the contained pigment, but their color is ordinarily concealed by the dense investiture of waxy powder and silken threads that characteris-

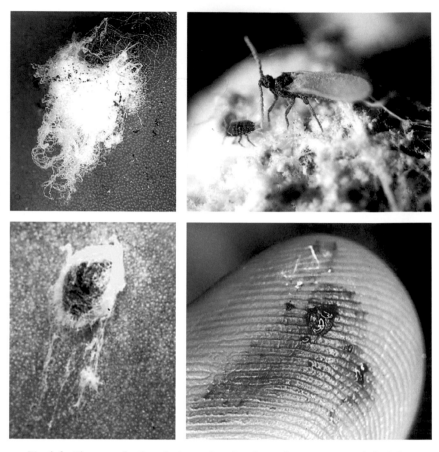

Top left: Close-up of a *Dactylopius confusus* female on *Opuntia*. Bottom left: A *D. confusus* female with the "woolly" cover partly removed, revealing the red coloring of the insect's body. Top right: A *D. confusus* male (winged individual) beside a newly hatched nymph. Bottom right: A female *D. confusus* killed by squashing, showing the deep red coloration of its contained tissues.

tically cloaks their bodies. The investiture is generally assumed to be defensive.

As to the function of carminic acid itself, none had been proposed until recently. This may seem surprising, given that the compound was so well known, but it is not unusual for the biological roles of familiar natural products to remain a mystery. However,

Top left: A larva of the pyralid moth, *Laetilia coccidivora,* hiding within the woolly investiture of the female cochineal bug it has consumed. Top right: A *L. coccidivora* adult. Bottom left: A larva of *L. coccidivora* regurgitating in response to being pinched with forceps. The caterpillar is arching its back and delivering a droplet of red vomit onto the offending instrument. Bottom right: An ant fleeing after attacking a *L. coccidivora* larva, leaving a trail in its wake as it attempts to wipe its body clean of the red oral effluent smeared upon its surface by the larva.

carminic acid has now been demonstrated to be a strong feeding deterrent to ants, which lends support to the notion that the compound is defensive.

Interestingly, *Dactylopius* are not without enemies. Three are known: a pyralid caterpillar *(Laetilia coccidivora),* a coccinellid beetle larva *(Hyperaspis trifurcata),* and a fly larva *(Leucopis* sp.). These are not only unaffected by carminic acid, but use the dye for defensive purposes of their own. *L. coccidivora* regurgitates the chemical

The pennyroyal stamp, issued by Britain in the 1800s. The pigment used in printing the stamp was carminic acid.

with crop fluid when disturbed, *H. trifurcata* emits it as part of droplets it exudes by reflex bleeding, and *Leucopis* sp. expels the compound with anal fluid.

The family Dactylopiidae includes only 9 species, all in the same genus.

REFERENCES

Donkin, R. A. 1977. Spanish red: an ethnological study of cochineal and the Opuntia cactus. *Transactions of the American Philosophical Society* 67:4–84.

Eisner, T., S. Nowicki, M. Goetz, and J. Meinwald. 1980. Red cochineal dye (carminic acid): its role in nature. *Science* 208:1039–1042.

Eisner, T., R. Ziegler, J. L. McCormick, M. Eisner, E. R. Hoebeke, and J. Meinwald. 1994. Defensive use of an acquired substance (carminic acid) by predaceous insect larvae. *Experientia* 50:610–615.

Fleming, S. 1983. The tale of cochineal: insect farming in the New World. *Archaeology* September/October:68–69, 79.

Takabayashi, J. 1993. Role of the scale wax of *Ceroplastes ceriferus* Anderson (Hemiptera: Coccidae) as a defense against the parasitic wasp *Anicetus ceroplastis* Ishii (Hymenoptera: Encyrtidae). *Journal of Insect Behavior* 6:107–115.

30

Class INSECTA
Order HEMIPTERA
Family Aleyrodidae
Metaleurodicus griseus
A whitefly

A *Metaleurodicus griseus* colony on one of its host plants
(*Eugenia* species).

 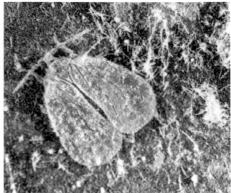

Left: Nymphs of *Metaleurodicus griseus.* Right: An adult *M. griseus,* resting on its colony's wax pad.

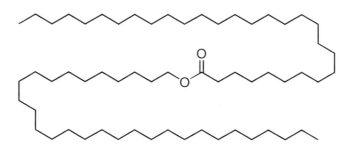

61. Triacontanyl triacontanoate

The tiny Aleyrodidae, or whiteflies, are often found in large aggregations composed of both immatures and adults. They are primarily tropical and subtropical in distribution, but include also pests of citrus and other cultivated trees, as well as of greenhouse and house plants. The nymphs lose their legs at the first molt and lead sedentary lives. The adult males and females, however, have functional legs and wings, and look like miniature moths. Colonies can usually be spotted by the white waxy material that the insects secrete and apply to the leaf surfaces they inhabit. *Metaleurodicus griseus,* for instance, a species native to Florida, forms conspicuous white aggregations on plants of the genus *Eugenia* (family Myrtaceae). In *M.*

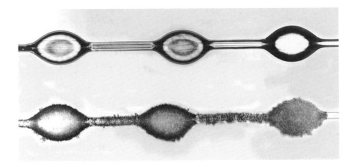

Top: Section of a normal viscid strand from the web of a spider
(Nephila clavipes). Bottom: A comparable section of strand bearing
the waxy powder left behind by a whitefly that brushed against the
strand.

griseus, the wax was shown to contain, in addition to a small amount
(1.2%) of long-chain unsaturated hydrocarbons, several long-chain
saturated esters, of which triacontanyl triacontanoate (**61**) makes up
69% of the mixture.

As with flatids (see Chapter 27), dactylopiids (see Chapter 29),
and woolly aphids (see Chapter 26), the secretory waxes of the
aleyrodids can be expected to serve in defense. For adult whiteflies, whose bodies are covered with a fine dusting of wax, the
powdery coating has been shown to prevent the insects from sticking to spider webs, which adult whiteflies must often encounter.
When whiteflies contact a web, instead of becoming entangled, they
merely lose some of their wax to the strands they touch, and fly on.
They are so small that their impact with the web may not even be
felt by the spider.

There are over 1,100 species in the family Aleyrodidae, assigned
to about 126 genera.

REFERENCES

Eisner, T., R. Alsop, and G. Ettershank. 1964. Adhesiveness of spider silk.
    Science 146:1058–1061.

Guerson, M., and D. Gerling. 2001. Parental care in the whitefly *Aleyrodes singularis. Ecological Entomology* 26:467–472.

Mason, R. T., H. M. Fales, M. Eisner, and T. Eisner. 1991. Wax of a whitefly and its utilization by a chrysopid larva. *Naturwissenschaften* 78:28–30.

Mound, L. H., and S. H. Halsey. 1978. *Whitefly of the World: A Systematic Catalogue of the Aleyrodidae (Homoptera) with Host Plant and Natural Enemy Data.* Chichester: British Museum/Wiley.

31

Class INSECTA
Order NEUROPTERA
Family Chrysopidae
Ceraeochrysa cubana
A green lacewing

Ceraeochrysa cubana.

62. Skatole

The Chrysopidae, or green lacewings, are among the most beautiful of the Neuroptera. Delicately built, they blend in with vegetation and are difficult to spot when at rest. They are primarily nocturnal, although they take to the wing both in the daytime and at night. They are attracted to lights and they are a common sight on screened windows.

Adult chrysopids have a number of defenses. When disturbed, they emit a stench from a secretion discharged from a pair of thoracic glands. One of the compounds in the secretion is skatole (62), well known as one of the odorous substances in mammalian feces. It is generally assumed that the secretion conveys protection by virtue of the stink.

When on the wing, chrysopids face danger from bats and spider webs. Both male and female chrysopids are acoustically sensitive to the echolocating pulses of hunting bats and are able therefore to detect approaching bats and take evasive action when pursued. They also have a special strategy for escaping from spider webs. They are so light that when they fly into an orb, they may not impart a strong enough jolt to the web to alert the spider. If the spider is feeding on a previous catch, it may ignore the chrysopid altogether. The chrysopid does not attempt to flutter free, but proceeds carefully to work itself out of the web. First, it uses its mouthparts to sever whichever strands are entangling its legs and antennae. When it then finds itself stuck by the wings only, it becomes still and lets gravity complete the job. Slowly, at an almost imperceptible rate, it slides downward over the web surface. It is able to do so because the hairs on its wings prevent the sticky strands of the web from making contact with the wing surface itself. It may take a *Ceraeochrysa cubana* more than an hour to slide free, but the strategy pays off.

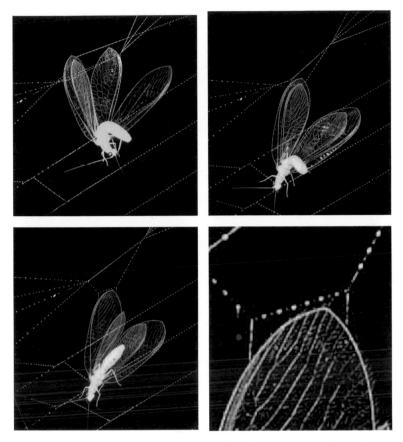

Ceraeochrysa cubana sliding out of a spider web. The pictures were taken at intervals of several minutes, and proceed from top left to top right, then to bottom left. At bottom left, the chrysopid is held by two stretched viscid droplets. The area of attachment is shown enlarged at bottom right. A moment after this picture was taken the chrysopid detached and fell free.

The skatole-containing defensive secretion, its stink notwithstanding, evidently offers no protection against orb-weaving spiders. In cases where an orb weaver does succeed in capturing a chrysopid, the spider usually makes a quick meal of it.

The Neuroptera, with only about 4,500 described species, account for only a small proportion of the insects. The Chrysopidae, with some 1,500 species in about 90 genera, form the largest family within the Neuroptera.

REFERENCES

Blum, M. S., J. B. Wallace, and H. M. Fales. 1973. Skatole and tridecene: identification and possible role in a chrysopid secretion. *Insect Biochemistry* 3:353–357.
Masters, W. M., and T. Eisner. 1990. The escape strategy of green lacewings from orb webs. *Journal of Insect Behavior* 3:143–157.
Miller, L. A. 1984. Hearing in green lacewings and their responses to the cries of bats. In M. Canard, Y. Séméria, and T. R. New, eds., *Biology of Chrysopidae.* The Hague: Junk.

32

Class INSECTA

Order NEUROPTERA

Family Chrysopidae

Ceraeochrysa smithi

A green lacewing

Left: The egg of an unidentified chrysopid, borne on a stalk devoid of oil droplets, as is typical for most chrysopid species. Middle: An egg of *Ceraeochrysa smithi*, on its oil-beset stalk. Right: A newly emerged larva of *C. smithi* descending the egg stalk and imbibing the oil droplets.

63. Oleic acid

64. Isopropyl myristate

65. Butanal

66. Decanal

67. Pentadecanal

Insects of the family Chrysopidae, the green lacewings, have the remarkable habit of laying their eggs on stalks. To produce an egg, the female first applies a droplet of clear gelatinous fluid from the tip of the abdomen to the substrate. She then flexes the abdomen abruptly upward, so as to pull the droplet into a thread and, after pausing briefly, squeezes out the egg. Thread-hardening occurs quickly, before the egg is entirely extruded. Chrysopids lay their eggs singly or in batches, and sometimes even in tight clusters with the stalks bundled together.

Ants are predators of chrysopids, and will if they can ascend a stalk, cut the egg free with their mandibles, and carry it away. One chrysopid, *Ceraeochrysa smithi,* has a way of keeping ants off the stalks. It coats the stalks with droplets of an oily secretion that is powerfully repellent to ants. Chemical analyses showed the fluid to consist of a mixture of long-chain fatty acids, including oleic acid (**63**); a fatty acid ester, isopropyl myristate (**64**); and a series of saturated aldehydes, including butanal (**65**), decanal (**66**), and pentadecanal (**67**).

The young larvae, when they emerge from the egg, have the problem of avoiding bodily contact with the oil when they descend along the stalk. They deal with the problem by ingesting the fluid. They pause by each droplet as they come down and suck it up with their hollow, sickle-shaped jaws. Upon arrival at the base of the stalk, they briskly walk away, to commence their life as hunters.

One other chrysopid species, *Leucochrysa floridana,* also coats its egg stalks with a deterrent fluid. Preliminary work has shown this fluid to differ chemically from that of *C. smithi.*

REFERENCES

Eisner, T., A. B. Attygalle, W. E. Conner, M. Eisner, E. MacLeod, and J. Meinwald. 1996. Chemical egg defense in a green lacewing (*Ceraeochrysa smithi*). *Proceedings of the National Academy of Sciences USA* 93:3280–3283.

33

Class INSECTA

Order NEUROPTERA

Family Chrysopidae

Chrysopa slossonae

A green lacewing

A *Chrysopa slossonae* larva among its aphid prey *(Prociphilus tessellatus)*.

The larvae of many green lacewings (family Chrysopidae) have the habit of covering themselves with debris. "Trash carriers" is the term applied to such larvae, in reference to the packet of inert matter they construct on their backs. Different species of chrysopids use differ-

Top left: A *Ceraeochrysa lineaticornis* larva bearing a packet of sycamore leaf tri-chomes. Top right: A *Leucochrysa pavida* larva with a packet made of lichen pieces. Bottom left: A larva of *Chrysopa slossonae* (arrow) nestled amidst its aphid prey. The guarding ant is imbibing a droplet of honeydew, unaware of the chrysopid's presence. Bottom right: A larva of *C. slossonae* under attack by an ant. The larva's wax packet was removed beforehand with forceps and the animal is now de-fenseless.

ent materials to fashion the packet. Some larvae use whatever debris they encounter in their meanderings, others use uneaten remnants of their insect prey, and still others use entirely different items. *Ceraeochrysa lineaticornis,* for instance, makes its packet exclusively from trichomes that it plucks from the leaves of sycamore trees. Another species, *Leucochrysa pavida,* uses primarily small pieces of lichens that it gathers from the bark of trees.

The packets are protective and probably serve primarily to deter ants. Ants are quickly discouraged if, upon attempting to bite into an insect, they succeed in grasping only inedible matter. Trash-carrying chrysopid larvae use the packet as a shield, which when attacked they maneuver so as to interpose it between themselves and the offending ants.

One chrysopid larva, *Chrysopa slossonae,* feeds exclusively on the woolly aphid, *Prociphilus tessellatus* (see Chapter 26), and constructs its package of the "wool"—the tufts of waxy filaments—that it plucks from the aphids. Covered by the wax, the larvae blend in with the aphids and are treated as aphids by the ants that guard the aphids. In a true sense, such larvae have taken on the role of "wolves in sheep's clothing," and the masquerade works. If such larvae are deprived of their acquired investiture, they are recognized and attacked by the ants. Cloaked in wool, they escape notice, and pass as aphids.

REFERENCES

Eisner, T., K. Hicks, M. Eisner, and D. S. Robson. 1978. "Wolf-in-sheep's-clothing" strategy of a predaceous insect larva. *Science* 199:790–794.

Eisner, T., J. E. Carrel, E. van Tassell, E. R. Hoebeke, and M. Eisner. 2002. Construction of a defensive trash packet from sycamore leaf trichomes by a chrysopid larva (Neuroptera: Chrysopidae). *Proceedings of the Entomological Society of Washington* 104:437–446.

Kennett, C. E. 1948. Defense mechanism exhibited by larvae of *Chrysopa californica* Coq. (Neuroptera: Chrysopidae). *Pan-Pacific Entomologist* 24:209–211.

LaMunyon, C. W., and P. A. Adams. 1987. Use and effect of an anal defensive secretion in larval Chrysopidae (Neuroptera). *Annals of the Entomological Society of America* 80:804–808.

34

Class INSECTA
Order COLEOPTERA
Family Carabidae
Galerita lecontei
A ground beetle

Galerita lecontei.

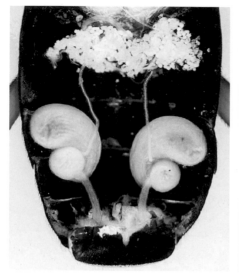

Left: Defensive glands of *Galerita lecontei* in a dissected preparation. Each gland consists of a cluster of secretory cells, a long, thin tubule that drains these and conveys the secretion to a bulbous kidney-shaped storage chamber, and a narrow ejaculatory duct. The storage chamber is enveloped in powerful compressor muscles. Right: A tethered *G. lecontei* discharging, first in response to pinching of a foreleg (top), and then in response to pinching of a hindleg (bottom). The spray pattern is depicted on phenolphthalein-impregnated filter paper.

Of the million or so insects that have been described, over one third are beetles. Beetles have succeeded in establishing themselves in virtually every terrestrial habitat, including the inland waters, and they truly represent one of the pinnacles of animal evolution. Beetles make up a well-characterized taxon, as was already recognized in the fourth century B.C.E. by Aristotle.

A distinguishing characteristic of beetles is the modification of the front wings into elytra, or wing covers. The elytra are characteristically rigid, fitting over the abdomen when at rest, with their inner edges in contact. Aside from providing protection for the membranous hindwings, which are folded beneath them, the elytra have made it possible for beetles to occupy enclosed spaces and cryptic habitats. When in flight, the elytra are usually held out at an angle, thus contributing to aerodynamic lift. The propulsive force is provided by the beat of the hindwings.

Although beetles derive advantages from the possession of the elytra, they are handicapped by the structures in one important way. For a beetle to take to the air, it must part the elytra before unfolding the hindwings. Escape is thus inevitably delayed. Not surprisingly, beetles have evolved a number of defenses that provide them with protection in emergencies. Most often these defenses take the form of dischargeable glands, present in the abdomen, and opening on or near the abdominal tip. Such glands are a standard feature in a number of beetle families, including the Carabidae, Dytiscidae, and Gyrinidae. Best known are the glands of the Carabidae. Those of *Galerita lecontei* provide a typical example.

G. lecontei is an agile beetle capable of running at great speed. It is a nocturnal hunter that feeds primarily on insects. Its defensive glands take the form of two identical organs, lying side by side in the abdominal cavity. Each gland consists of a tight cluster of some 250 secretory cells, arranged like grapes in a bunch, and connected by way of drainage tubules to a main efferent duct that leads to the storage chamber of the gland. This duct is lengthy and ordinarily coiled. The storage chamber is enveloped by compressor muscles, and empties to the outside by way of a short ejaculatory duct, which terminates to the side of the anus, opposite its counterpart from the other gland.

A *G. lecontei* that has gone unmolested for some weeks can discharge 6 or 7 times when persistently disturbed. The secretion is potently irritating and has a distinct acidic odor. Chemical analysis showed the fluid to consist of a mixture of formic acid (**68**), acetic acid (**1**), and a number of lipophilic compounds, principally nonane (**69**) and decyl acetate (**70**). At a concentration of 80%, formic acid is the dominant component. The other constituents make up another 10% of the mixture, and the remainder is water.

When replete, the glands of *G. lecontei* contain about 5 milligrams of secretion, equal to about 3% of the beetle's body mass. After 11 days, beetles that had discharged their entire glandular contents had produced about 1.4 milligrams of formic acid, roughly 35% of the total glandular load. Although the secretion is replen-

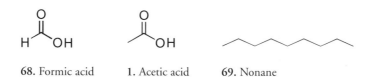

68. Formic acid   **1.** Acetic acid   **69.** Nonane

70. Decyl acetate

ished rather slowly in terms of days, individual gland cells have a relatively high secretory output. Since the beetle has only 500 gland cells, a production of 1.4 milligrams formic acid per 11 days is indicative of a secretory rate of 10 nanograms per cell per hour, which is about twice the rate of secretion of hydrochloric acid by the parietal cells of the mammalian stomach.

By flexing the abdominal tip, *G. lecontei* is able to aim its discharges in different directions. This can be demonstrated by tethering the beetle and causing it to spray on filter paper impregnated with an alkaline solution of phenolphthalein. Wherever hit by the spray, the paper turns from red to white. Pinching a beetle on a leg with forceps causes the animal to discharge accurately toward that leg. The spray mechanism is evidently designed to be used against small enemies such as ants, which might be missed unless the discharges are directed. Ants, in fact, are likely to be among the beetle's chief enemies, although spiders, rodents, and birds could be real threats as well.

Formic acid, at a concentration of 80%, is a formidable irritant, which raises the question of how the beetle is able to produce the substance without injury to itself. The answer may lie in the fact that the glands of the beetle are lined throughout with a thin but impervious cuticular membrane, essentially a membranous extension of the outer cuticle. Even the gland cells themselves have an inner cuticular apparatus, consisting of a series of convergent tubules, by which the secretion is presumably routed from its site of production within the cells to the efferent duct that conveys it to the stor-

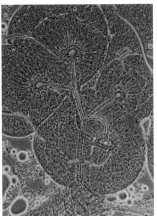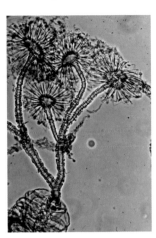

Left: Cells from the cluster of secretory tissue of a *Galerita lecontei* defensive gland. The cells are connected by fine ducts that eventually merge to form the drainage tubule that leads to the storage chamber of the gland. Middle: A cluster of secretory cells, showing the branched inner ducts, each ending in a collecting apparatus, into which the glandular fluid is presumably secreted. Right: The cellular duct system, complete with collecting apparatuses, isolated by treatment of a group of secretory cells with potassium hydroxide.

age chamber. Such intracellular cuticular tubule systems are quite consistently found in the secretory cells of arthropodan defensive glands, and may explain why so many of these animals are able to produce powerful toxicants without poisoning themselves.

Formic acid is produced by a number of carabid beetles and is also the primary component in the spray of ants of the subfamily Formicinae (see Chapter 68). Such ants doubtless figure among *G. lecontei*'s enemies, and one cannot help wondering whether they are tolerant of the beetle's spray. It is conceivable that the lipophilic components of the beetle's secretion, which may themselves be deterrent but absent from the ant's secretion, impart a "mark of distinction" to the fluid, such that the mixture is rendered effective against formic acid–producing ants.

It is likely that formic acid is biosynthesized by *G. lecontei,* as it is in ants, from L-serine and glycine. As a carnivore, *G. lecontei* is sure to obtain these precursors with its diet. If we assume, somewhat arbitrarily, that the beetle obtains 5% each of L-serine and glycine with its food, then 60 milligrams of protein would be needed by the beetle to load its glands to capacity. For the beetle to synthesize the amount of formic acid (0.7 milligrams) that it expends in a single discharge would require that it invest 9 milligrams of protein in formic acid production. At the usual rate at which it secretes the acid, that would amount to 1.6 milligrams of protein per day, an equivalent of 1% of the beetle's mass. Defense in *G. lecontei,* as in *Homo sapiens,* evidently comes at a cost.

The Carabidae make up a large family, with some 30,000 species in about 1,500 genera. They are of worldwide distribution.

REFERENCES

Dazzini-Valcurone, M., and M. Pavan. 1980. Glandole pigidiali e secrezioni difensive dei Carabidae (Insecta Coleoptera). *Pubblicazione Istituto di Entomologia della Università di Pavia* 12:1–36.

Forsyth, D. J. 1972. The structure of the pygidial defence glands of Carabidae (Coleoptera). *Transactions of the Zoological Society, London* 32:249–309.

Moore, B. P. 1979. Chemical defense in carabids and its bearing on phylogeny. In T. L. Erwin, G. E. Ball, and D. R. Whitehead, eds., *Carabid Beetles: Their Evolution, Natural History, and Classification.* The Hague: Junk.

Rossini, C., A. B. Attygalle, A. González, S. R. Smedley, M. Eisner, J. Meinwald, and T. Eisner. 1997. Defensive production of formic acid (80%) by a carabid beetle *(Galerita lecontei). Proceedings of the National Academy of Sciences USA* 94:6792–6797.

Schildknecht, H., U. Maschwitz, and H. Winkler. 1968. Zur Evolution der Carabiden-Wehrdrüsensekrete. *Naturwissenschaften* 55:112–117.

Will, K. W., A. B. Attygalle, and K. Herath. 2000. New defensive chemical data for ground beetles (Coleoptera: Carabidae): interpretations in a phylogenetic framework. *Biological Journal of the Linnean Society* 71:459–481.

35

Class INSECTA

Order COLEOPTERA

Family Carabidae

Brachinus (many species)

Bombardier beetles

A bombardier beetle (an unidentified South American species).

Most remarkable among the Carabidae, certainly as regards possession of defenses, are the bombardier beetles. When seized, these insects emit a popping sound at the same time that they eject a spray from the rear end. The discharge feels hot, and under the proper lighting conditions may be visible as a faint cloud.

Bombardier beetles occur throughout much of the world, in habitats as diverse as forests, plains, and deserts. They belong to several genera. The North American species are all of the genus *Brachinus,*

16. 2-Methyl-1,4-benzoquinone   **30.** 1,4-Benzoquinone

H₂O₂

71. Hydrogen peroxide   **72.** 2-Methylhydroquinone

and although they come in a range of sizes, look very much alike. Their wing covers are typically iridescent blue, while their body and legs are reddish-brown. African and South American bombardiers, such as those of the genera *Stenaptinus* and *Pheropsophus,* are larger than *Brachinus,* and often have yellowish markings on the elytra. The burning sensation imparted by these larger bombardiers when you take them in your hand is quite startling and usually results in the beetle being flung away. Even the popping sounds themselves, which accompany the emissions, can be disconcerting.

The spray apparatus of bombardier beetles is well understood. It varies little in the different species of bombardiers. The chemicals discharged by the beetles are 1,4-benzoquinones (for example, **16** and **30**). These are potent irritants even when cold, as is indicated explicitly on the labels of the containers in which these compounds are marketed. Many arthropods have independently evolved the ability to produce 1–4-benzoquinones for defense, including, among others, millipedes, earwigs, termites, grasshoppers, and cockroaches (see Chapters 8, 15–17, and 21).

What is remarkable about bombardier beetles is that they do not store their benzoquinones as such in the glands, but produce the

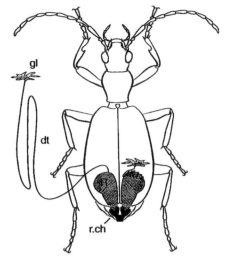

gl

dt

r.ch

Diagram of the defensive apparatus of a bombardier beetle. The tip of the abdomen, bearing the gland openings, is fully revolvable, which enables the beetle to aim its spray in all directions. Each gland consists of a cluster of secretory tissue (gl), a duct (dt) that drains this tissue, and a compressible, muscled storage chamber—the reservoir (R). Interposd between the reservoir and the gland opening is the reaction chamber (r.ch). The reservoir stores the reactants (hydrogen peroxide, hydroquinones), which are forced into the reaction chamber when the beetle is disturbed, there to initiate the enzyme-catalyzed events that lead to quinone formation and the discharge.

chemicals by explosive synthesis at the moment of ejection. Their defensive glands are essentially binary weapons. They are two-chambered, unlike carabid glands generally, and constructed in such a way that the contents of the two chambers are mixed during ejection. The two chambers differ greatly in size. The larger inner chamber, or reservoir, contains a mixture of hydrogen peroxide (71) and hydroquinones (for example, 72). The smaller chamber, through which the contents of the reservoir must pass upon being discharged, contains two kinds of enzymes: catalases and peroxidases. There is a valve between the two chambers that is ordinarily kept closed, keeping the reservoir contents from leaking into the smaller chamber. The smaller chamber is essentially a reaction chamber. When the beetle is disturbed, as when one of its legs is bitten by an ant, it responds promptly by activating the compressor muscles surrounding the reservoir. This causes the reservoir contents to be forced into the reaction chamber, where the catalases promote the liberation of oxygen from hydrogen peroxide, and the peroxidases catalyze the oxidation of the hydroquinones to quinones by use of the oxygen freed from the hydrogen peroxide. Simultaneously, under pressure from the liberated oxygen, the mixture pops out.

Left: A tethered bombardier beetle responding to the pinching of the left foreleg by discharging forward. Right: A tethered bombardier beetle discharging in response to an ant attack. The ant was also tethered, and held so it would bite the left hindleg of the beetle. The beetle responded instantly by spraying.

Thermodynamic calculations predicted that enough heat should be liberated as a consequence of the chemical events in the reaction chamber to bring the expelled mixture to the boiling point. Because the reactants are in aqueous solution, this means a temperature of 100°C. Actual measurements of the spray temperature, made by causing the beetle to discharge upon an electronic thermometer (a thermistor of known conductance), confirmed this prediction. Bombardier beetles do indeed eject their spray at the temperature of boiling water!

Bombardier beetles are extraordinary marksmen. The fact that they eject their secretion with an audible popping sound made it possible to photograph them at the very moment that they sprayed. The technique involved photographing the beetles by the light of an electronic flash unit, hooked up to a microphone that picked up the sound of the ejections and triggered the flash. The photos showed that the beetles can spray in virtually any direction, and that they can accurately target any part of their body subjected to assault. The spray is effective against ants, spiders, frogs, and birds. But orb-weaving spiders that wrap their prey in silk before inflicting their lethal bites may avoid getting sprayed. They wrap the bombardiers gently, usually without inducing them to discharge, and when they finally bear down to bite, the beetles may be too tightly enveloped to revolve their rear ends and take aim. If the beetles do discharge, the spiders are only temporarily disturbed. The beetle, once spun in, is usually doomed.

Acoustical recordings of the discharge sounds suggested that the beetle's ejections are pulsed, like the fluid delivery of a dental water pick. High-speed filming confirmed this. The beetle ejects its spray in squirts, at the rate of 500 to 1,000 squirts per second. This has been interpreted to indicate that the chemical events in the reaction chamber are discontinuous, proceeding essentially as a quick series of microexplosions. By pulsing the delivery of its spray, the beetle is able to exercise greater control over the rate of output of its defensive fluid, while also preventing the reaction chamber from overheating.

Bombardier beetles are probably among the best defended of animals. They are remarkably long-lived as adults. Individuals of some of the larger tropical species have survived for as long as several years in captivity. Mortality is probably high among bombardier beetle larvae. These have complicated trophic requirements—some parasitize aquatic beetles, others feed on mole cricket eggs—and such finicky dependencies are sometimes associated with low survivorship.

REFERENCES

Aneshansley, D. J., T. Eisner, J. M. Widom, and B. Widom. 1969. Biochemistry at 100°C: explosive secretory discharge of bombardier beetles *(Brachinus). Science* 165:61–63.

Dean, J. 1979. Defensive reaction time of bombardier beetles: an investigation of the speed of a chemical defense. *Journal of Chemical Ecology* 5:691–701.

———— 1980. Encounters between bombardier beetles and two species of toads *(Bufo americanus, Bufo marinus):* speed of prey-capture does not determine success. *Journal of Comparative Physiology* 135:41–50.

———— 1980. Effect of thermal and chemical components of bombardier beetle chemical defense: glossopharyngeal response in two species of toads *(Bufo americanus, Bufo marinus). Journal of Comparative Physiology* 135:51–59.

Dean, J., D. J. Aneshansley, H. E. Edgerton, and T. Eisner. 1990. Defensive spray of the bombardier beetle: a biological pulse jet. *Science* 248:1219–1221.

Eisner, T., and D. J. Aneshansley. 1999. Spray aiming in the bombardier beetle: photographic evidence. *Proceedings of the National Academy of Sciences USA* 96:9705–9709.

Schildknecht, H. 1957. Zur Chemie des Bombardierkäfers. *Angewandte Chemie* 69:62.

Schildknecht, H., and K. Holoubek. 1961. Die Bombardierkäfer und ihre Explosionschemie. V. Mitteilung über Insekten-Abwehrstoffe. *Angewandte Chemie* 73:1–7.

36

Class INSECTA

Order COLEOPTERA

Family Gyrinidae

Dineutus hornii

A whirligig beetle

Dineutus hornii at rest on the water's surface.

73. Gyrinidal

The aquatic beetles of the family Gyrinidae are commonly known as whirligigs, in reference to their habit of swimming in tight circles or gyrating on the surface of the water. Gyrinids are streamlined like dytiscids (see Chapter 37), but are readily recognized by the fact that their eyes are divided into an upper and lower portion, the upper portion for vision in air, the lower portion for vision in water. Gyrinids commonly aggregate to form swarms, and are a familiar sight in ponds and streams. The adults prey on organisms that drop into the water. The larvae are also predacious, but are bottom feeders. Gyrinids are worldwide in distribution. The family contains 11 genera and some 700 species.

Silhouetted against the sky, as they swim on the surface of the water, adult gyrinids must be easily detectable by fish. The beetles can risk such a life style because they are chemically protected. Like Carabidae (see Chapters 34 and 35) and Dytiscidae (see Chapter 37), they possess a pair of defensive glands, opening on the abdominal tip, that they put to use when they are attacked by fish. In species of two genera of gyrinids, *Dineutus* and *Gyrinus,* the principal product of these glands is the highly oxidized norsesquiterpene gyrinidal (73).

Experiments with largemouth bass *(Micropterus salmoides)* and *Dineutus hornii* showed the fish to reject this gyrinid. Moreover, the bass also rejected edible food items (mealworms) that were coated with gyrinidal. The basses' oral tolerance of gyrinidal varied broadly, depending on the fishes' state of satiation. When the bass were hungry, they sometimes accepted mealworms treated with over 100 micrograms of gryrinidal, the full quantity ordinarily contained in the glands of *D. hornii.* When the fish were close to satiation, they

A specimen of *Dineutus hornii* that has been treated with potassium hydroxide to reveal the two defensive glands present in the rear of its abdomen.

frequently rejected mealworms bearing as little as 0.5 micrograms gyrinidal.

The fish did not always reject the treated food items outright. Instead, they subjected the items to an intensive cleansing behavior, involving what has been called oral flushing. This behavior is of abrupt onset and quite typical. As soon as a fish takes a noxious item in the mouth it begins a rhythmic opening and closing of the mouth, at the same time that it opens and closes its opercular flaps (the gill covers). Mouth and flaps open in alternation, with the result that water is continuously pumped in and out of the oral cavity. The food item is held in the mouth during this behavior and is rinsed as a result. The item is sometimes spat out briefly in the course of rinsing, but is then taken in again for further flushing, and is eventually either swallowed or definitively rejected. The length of

Frames from a motion picture film showing a largemouth bass *(Micropterus salmoides)* taking, and then rejecting, a gyrinidal-bearing mealworm.

time that the fish invests in this behavior is a function of its hunger and the amount of gyrinidal in the food item. The hungrier the fish, and the greater the amount of gyrinidal, the longer the flushing period. Quite clearly, flushing is intended to rid food items of their noxious taste.

The beetle is adapted to cope with the basses' flushing behavior. Unlike terrestrial beetles such as carabids, which eject their secretion in short bursts, *D. hornii* doles out its secretion in a slow trickle. The beetle has the capacity to deliver secretion over a period averaging about 1.5 minutes. Flushing time for a half-satiated bass is on the order of 1 minute. The beetles therefore have a good chance of being released when caught by bass.

It is now known that gyrinids produce other norsesquiterpenes, similar to gyrinidal, in their glands. These compounds are most probably also defensive.

REFERENCES

Dettner, K. 1987. Chemosystematics and evolution of beetle chemical defenses. *Annual Review of Entomology* 32:17 48.

Dettner, K., and G. Schwinger. 1980. Defensive substances from pygidial glands of water beetles. *Biochemical Systematics and Ecology* 8:89–95.

Eisner, T., and D. J. Aneshansley. 2000. Chemical defense: aquatic beetle (*Dineutes hornii*) vs. fish *(Micropterus salmoides). Proceedings of the National Academy of Sciences USA* 97:11313–11318.

Forsyth, D. J. 1968. The structure of the defence glands in the Dytiscidae, Noteridae, Halipidae, and Gyrinidae (Coleoptera). *Transactions of the Royal Society of London* 120:159–182.

Meinwald, J., K. Opheim, and T. Eisner. 1972. Gyrinidal: A sesquiterpenoid aldehyde from the defensive glands of gyrinid beetles. *Proceedings of the National Academy of Sciences USA* 69:1208–1210.

——— 1973. Chemical defense mechanisms of arthropods. 36. Stereospecific synthesis of gyrinidal, a norsesquiterpenoid aldehyde from gyrinid beetles. *Tetrahedron Letters* 4:281–284.

Miller, J. R., L. B. Hendry, and R. O. Mumma. 1975. Norsesquiterpenes as defensive toxins of whirligig beetles (Coleoptera Gyrinidae). *Journal of Chemical Ecology* 1:59–82.

Schildknecht, H. 1970. The defensive chemistry of land and water beetles. *Angewandte Chemie—International Edition in English* 9:1–19.

37

Class INSECTA

Order COLEOPTERA

Family Dytiscidae

Thermonectus marmoratus

A predaceous diving beetle

Thermonectus marmoratus.

The dytiscids, or predacious diving beetles, are familiar inhabitants of ponds, streams, and the edges of lakes. They are good swimmers and favor shallow waters, where they prey on virtually any animals they can catch. Some dytiscids, such as those of the genera *Dytiscus* and *Cybister,* are large enough to be able to feed on small fish. Further prey includes tadpoles, leeches, snails, and aquatic insect larvae.

Thermonectus marmoratus responding to being held in fingers by discharging pale milky secretion from its prothoracic glands. The fluid has spread over the pronotum of the beetle.

The adults are good flyers and regularly take to the air as they migrate from one aquatic site to another. The larvae are also aquatic and predacious. The family includes some 120 genera and 3,000 species.

Like the Gyrinidae (see Chapter 36), dytiscids are chemically protected. They have a pair of defensive glands, opening just behind the head on the anterolateral margins of the prothorax, from which they eject a milky fluid when disturbed. The principal constituents of this secretion are steroids (pregnanes) that have been shown to be both distasteful and toxic to fish and amphibians. In one particularly beautiful dytiscid, the speckled *Thermonectus marmoratus,* the secretion contains two steroids, cybisterone (74) and mirasorvone (75). The former compound occurs in other dytiscids as well, but mirasorvone, an 18-oxygenated steroid, is not known from other insects. Interestingly, one other group of aquatic insects, the belostomatid bugs (see Chapter 24), also produces steroids for defense against fish.

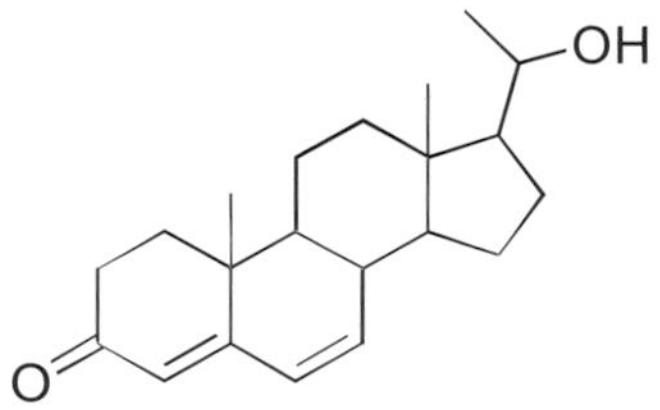

74. Cybisterone

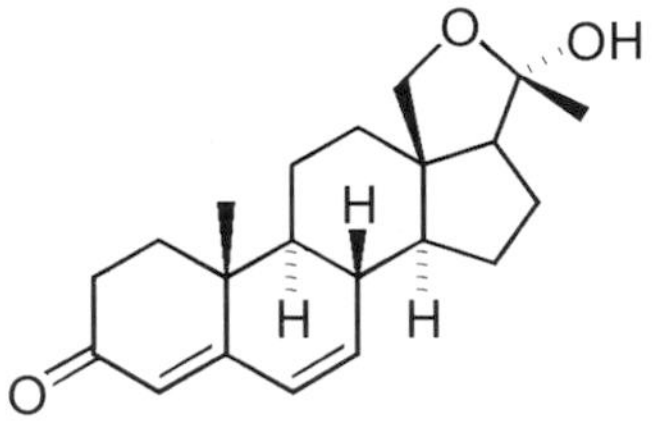

75. Mirasorvone

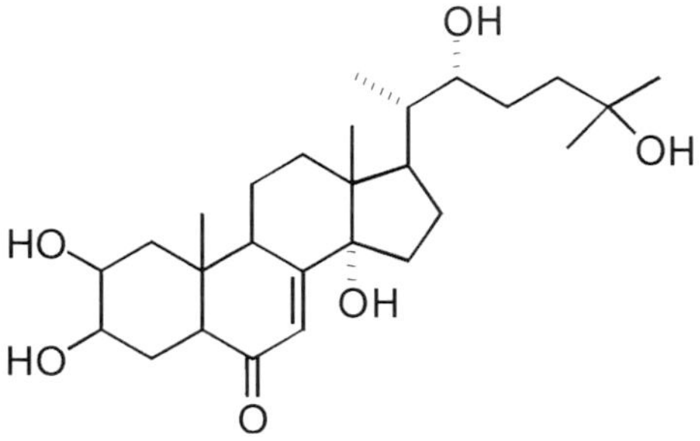

76. α-Ecdysone

Steroids in insects include compounds that function as hormones or pheromones. One of the principal hormones regulating insect growth, α-ecdysone (76), is a steroid, as are the pheromones that control aggregation behavior in the cockroach *Blatella germanica.*

The use of steroids for defense is not restricted to aquatic insects. Fireflies (see Chapter 41) and silphid beetles of the genus *Silpha* (see Chapter 38) are protected by steroids, as is the monarch butterfly (see Chapter 66), which incorporates defensive steroids from its larval foodplant.

Dytiscids have a second pair of glands, besides the thoracic pair, opening at the tip of the abdomen, from which they discharge a mixture of compounds, including benzoic acid (15) and methyl *p*-hydroxybenzoate (77). The possibility has not been ruled out that these glandular products confer protection against predators, but it has also been suggested that the beetles employ the substances as topical disinfectants. Indeed, dytiscids have been observed to anoint

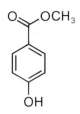

15. Benzoic acid   77. Methyl *p*-hydroxybenzoate

themselves with the secretion from their abdominal glands, using their legs to apply the material to their body surfaces when they are at rest on land.

REFERENCES

Blunck, H. 1917. Die Schreckdrüsen der *Dytiscus* und ihr Sekret. *Zeitschrift für Wissenschaftliche Zoologie* 117:205–256.

Dettner, K. 1987. Chemosystematics and evolution of beetle chemical defenses. *Annual Review of Entomology* 32:17–48.

Dettner, K., and G. Schwinger. 1980. Defensive substances from pygidial glands of water beetles. *Biochemical Systematics and Ecology* 8:89–95.

Fescemyer, H. W., and R. O. Mumma. 1983. Regeneration and biosynthesis of dytiscid defensive agents (Coleoptera: Dytiscidae). *Journal of Chemical Ecology* 9:1449–1464.

Forsyth, D. J. 1968. The structure of the defence glands in the Dytiscidae, Noteridae, Halipidae, and Gyrinidae (Coleoptera). *Transactions of the Royal Society of London* 120:159–182.

Gerhart, D. J., M. E. Bondura, and J. A. Commito. 1991. Inhibition of sunfish feeding by defensive steroids from aquatic beetles: structure-activity relationships. *Journal of Chemical Ecology* 17:1363–1370.

Maschwitz, U. 1967. Eine neuartige Form der Abwehr von Mikroorganismen bei Insekten. *Naturwissenschaften* 54:649.

Meinwald, J., Q. Huang, J. Vrkoc, K. B. Herath, Z.-C. Yang, F. Schröder, A. B. Attygalle, V. K. Iyengar, R. C. Morgan, and T. Eisner. 1998. Mirasorvone: a masked 20-ketopregnane from the defensive secretion of a diving beetle *(Thermonectus marmoratus)*. *Proceedings of the National Academy of Sciences USA* 95:2733–2737.

Schaaf, O., J. Baumgarten, and K. Dettner. 2000. Identification and function of prothoracic exocrine gland steroids of the dytiscid beetles *Graphoderus cinereus* and *Laccophilus minutus. Journal of Chemical Ecology* 26:2291–2305.

Schildknecht, H. 1970. The defensive chemistry of land and water beetles. *Angewandte Chemie—International Edition in English* 9:1–19.

Schildknechht, H., and K. H. Weis. 1962. Zur Kenntniss der Pygidialblasensubstanzen vom Gelbrandkäfer (*Dytiscus marginalis* L.) XIII. Mitteilung über Insektenabwehrstoffe. *Zeitschrift für Naturforschung* 17b:448–452.

Schildknecht, H., D. Hotz, and U. Maschwitz. 1967. Über Arthropoden-Abwehrstoffe. XXVII. Die C_{21}-Steroide der Prothorakalwehrdrüsen von *Acilius sulcatus. Zeitschrift für Naturforschung* 22b: 938–944.

38

Class INSECTA
Order COLEOPTERA
Family Silphidae
Necrodes surinamensis
The red-lined carrion beetle

Necrodes surinamensis.

Beetles are extraordinarily diverse in their feeding habits. There is hardly a food source that is not exploited by these insects. Of some interest are the beetles of the family Silphidae, constituting some 14 genera and 175 species, many of which are attracted to carrion. They compete for this resource with many other insects, notably flies.

Silphid adults have evolved a number of chemical defenses. *Necrodes surinamensis* relics on a single gland, consisting of an elongate tubular structure that conveys its products to a pouch connected to

2. Caprylic acid

78. Capric acid

79. *(E)*-3-Decenoic acid

80. *(E)*-4-Decenoic acid

81. Lavandulol

82. α-Necrodol

83. β-Necrodol

the rectum. The gland secretes a complex chemical mixture, which the beetle ejects as a spray through the anus. The tip of the abdomen is flexible in *N. surinamensis,* and so the beetle can aim its discharges in any direction. The components of the secretion are a series of fatty acids—caprylic acid (**2**), capric acid (**78**), *(E)*-3-decenoic acid (**79**), and *(E)*-4-decenoic acid (**80**)—lavandulol (**81**), and two previously unknown terpenes, α-necrodol (**82**) and β-necrodol (**83**). As a consequence of its lavandulol content, the secretion is pleasantly scented. But all compounds of the mixture, including lavandulol, have the potential to deter enemies. Tests with ants *(Formica exsectoides)* and thrushes *(Catharus ustulatus)* showed that these predators reject *N. surinamensis.*

The necrodols are compounds of broad anti-insectan activity. If a glass capillary tube bearing β-necrodol is held beside individual insects that have alighted on an illuminated sheet at night, a large fraction, including beetles, hymenopterans, and hemipterans, will be prompted to walk or fly away. Interestingly, α-necrodol has now been found to occur (together with lavandulol) as part of the defensive oil of a mint plant *(Lavandula luisieri).*

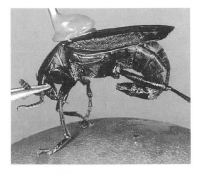

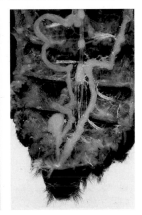

Top left: A tethered *Necrodes surinamensis,* pointing its abdominal tip forward beneath the body to take aim at the left foreleg, which is being pinched with forceps. Top right: The same beetle taking aim at a warm spatula that is being applied to its back. Bottom left: The beetle ejecting its spray toward its right midleg. The spray pattern is revealed on phenolphthalein-impregnated indicator paper. Bottom right: The abdomen of *N. surinamensis,* laid open to show the hindgut (the slender white tube on the right) and the associated defensive gland, consisting of a glandular tube (on the left) and a storage sac (at lower left).

Silphid beetles of the genus *Silpha* had long been known to emit a malodorous ooze from the anus when disturbed. The effluent was shown to be rich in ammonia, which could in itself render such beetles unacceptable. But the anal fluid of *Silpha* may owe its deterency to other compounds as well. In *Silpha americana,* the ooze contains a number of steroids (pregnanes), principally 15-β-hydroxy-progesterone (**84**), none previously known from insects. Tests with

84. 15β-Hydroxyprogesterone

85. 3α,7β,14β-Pregn-4-ene-15,20-dione

86. 3α,7β,20-Trihydroxy-14β-pregn-4-ene-15-one

two of the steroids (including **84**) showed these compounds to deter jumping spiders, although they are doubtless effective also against other predators. Another species of *Silpha, S. novaboracensis,* has been shown to produce two additional pregnanes (**85, 86**), not previously known from natural sources. Beetles of the genus *Silpha* produce their steroids in a gland attached to the rectum, but this gland differs fundamentally in structure from that of *N. surinamensis,* and may have evolved independently.

REFERENCES

Eisner, T., and J. Meinwald. 1982. Defensive spray mechanism of a silphid beetle *(Necrodes surinamensis). Psyche* 89:357–367.

Eisner, T., M. Deyrup, R. Jacobs, and J. Meinwald. 1986. Necrodols: anti-insectan terpenes from defensive secretion of carrion beetle *(Necrodes surinamensis). Journal of Chemical Ecology* 12:1407–1415.

Garcia-Vallejo, M. I., M. C. Garcia-Vallejo, J. Sanz, M. Bernabe, and A. Velasco-Negueruela. 1994. Necrodane (1,2,2,3,4-

pentamethylcyclopentane) derivatives in *Lavandula luisieri,* new compounds to the plant kingdom. *Phytochemistry* 36:43–45.

Meinwald, J., B. Roach, and T. Eisner. 1986. Defensive steroids from a carrion beetle *(Silpha novaboracensis). Journal of Chemical Ecology* 13:35–38.

Meinwald, J., B. Roach, K. Hicks, D. Alsop, and T. Eisner. 1985. Defensive steroids from a carrion beetle *(Silpha americana). Experientia* 41:516–519.

Roach, B., T. Eisner, and J. Meinwald. 1990. Defense mechanisms of arthropods. 83: α and ß-necrodol, novel terpenes from a carrion beetle. (*Necrodes surinamensis,* Silphidae, Coleoptera). *Journal of Organic Chemistry* 55:4047–4051.

Schildknecht, H., and K. H. Weiss. 1962. Über die chemische Abwehr der Aaskäfer. XIV. Mitteilung über Insektenabwehrstoffe. *Zeitschrift für Naturforschung* 17b:452–455.

39

Class INSECTA
Order COLEOPTERA
Family Staphylinidae
Creophilus maxillosus
The hairy rove beetle

Creophilus maxillosus.

The Staphylinidae, or rove beetles, make up a large family, containing some 1,500 genera and more than 25,000 species. A rove beetle's body is usually elongate and narrow, and its elytra are often short, so that much of the abdomen is exposed. The rear of the abdomen is flexible and is often pointed upward when the animal is disturbed.

Adult staphylinids occur in a great diversity of habitats. They are found beneath stones, along ponds and streams, in leaf litter, in the soil, under bark, in manure and carrion, in termite and ant colonies,

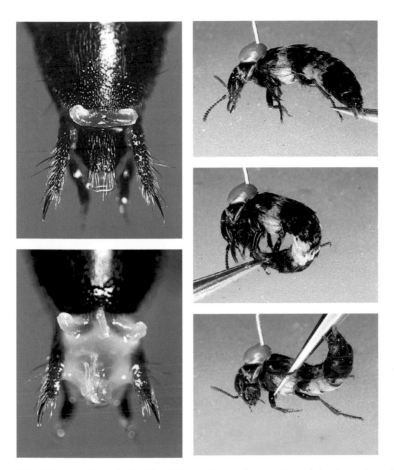

Creophilus maxillosus. Left: The abdominal tip, after eversion of glands (top), and a moment later, after emission of rectal fluid (bottom). Right: A tethered beetle revolving its abdominal tip in response to being "bitten" with forceps.

and in caves. Most staphylinids are predacious, but some feed on fungi, fungal spores, or algae. They walk about rapidly and are often quick to take to the air when startled.

Defensive glands, present in the abdomen and opening at the abdominal tip, are probably ubiquitous in rove beetles. The products of these glands have been examined chemically in a number of species. It is clear that rove beetles produce defensive substances of considerable diversity.

Creophilus maxillosus is a typical example. It is one of the largest North American staphylinids and is commonly found in carrion. The beetle's glands are two cuticular sacs, ordinarily kept involuted dorsally between abdominal segments 7 and 8. The beetle everts the glands promptly when disturbed, but the disturbance has to be forcible. Mild manipulation does not usually induce eversion. Only if the animal is pinched, say by forceps, as an ant might do with its mandibles, does it evert its glands in full. The everted sacs are visibly wet with secretion.

When it everts the glands, the beetle usually defecates as well. It protrudes the anal tube and squeezes out a droplet of fluid, which spreads over the everted glands and becomes mixed with the secretion. Thus "armed," with its abdominal tip wetted, the beetle flexes its abdomen so as to bring the tip to the site of trouble. The abdomen is extraordinarily flexible. It can be rotated upward at a sharp angle, as well as beneath the body, to almost within reach of the head. There is therefore virtually no part of the body surface that the beetle cannot wet with defensive fluid. Experiments with ants showed that they are effectively deterred by the beetle. In warding off the ants, the beetle relies not only on its glands but on its mandibles, which are powerful enough to decapitate an ant or cut it in half.

Chemical analysis showed that the secretion of *C. maxillosus* contains six components: isoamyl alcohol (**87**), isoamyl acetate (**88**), iridodial (**89**), actinidine (**90**), dihydronepetalactone (**91**), and *(E)*-8-oxocitronellyl acetate (**92**). Tests with ants showed that most of the major components, and especially the chief constituent (**91**), are effectively repellent.

Some components of the *C. maxillosus* secretion (**89, 90**) are produced also by other rove beetles. Component **91** is closely related to compounds such as anisomorphal (**43**) and nepetalactone (**44**), known from a number of insectan and plant sources (see Chapter 20). Compound **88** is well known as a component of the alarm pheromone of honey bees (see Chapter 69). Such parallel production of the same or similar substances by organisms of diverse kinds is not unusual, given the overlap in biosynthetic capacities of organisms generally.

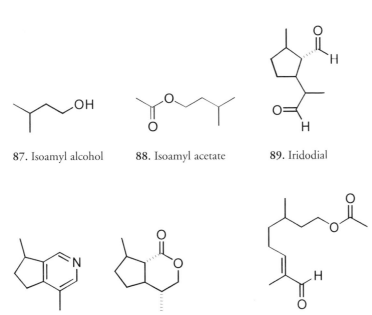

87. Isoamyl alcohol   **88.** Isoamyl acetate   **89.** Iridodial

90. Actinidine   **91.** Dihydronepetalactone   **92.** *(E)*-8-Oxocitronellyl acetate

43. Anisomorphal   **44.** Nepetalactone

Quite remarkable is the use to which the defensive glands are put in staphylinds of the genus *Stenus*. These beetles are terrestrial, but they occasionally forage on water or find themselves blown onto water by the wind. If they are then attacked, they resort to an escape behavior that has been named appropriately by its German discoverers: *Entspannungsschwimmen*—quite literally "swimming by depression of surface tension". What the beetles do is touch the tip of the abdomen to the water, causing the surface tension to be depressed behind them, so that they are then propelled forward as the surface recedes before them. The velocity they achieve is considerable. Dis-

An unidentified rove beetle of the genus *Paederus*.

93. Pederin

tances of up to 15 meters may be covered in one stretch, at a speed of up to 75 centimeters per second. The beetles resort to such escape behavior only in emergencies, as possibly when pursued by water striders (Hemiptera of the family Gerridae).

One of the most interesting compounds ever obtained from insects is pederin (**93**), a product of staphylinids of the genus *Paede-*

rus. Pederin is an irritant and vesicant, active in vanishingly small amounts. In regions where *Paederus* occasionally swarm, the beetles can be the source of serious outbreaks of eye inflammation. Pederin is present in the female beetle only, and is apparently synthesized by microbial symbionts. Compounds related to pederin have been isolated from marine sponges and seem to be produced in these primitive animals by microorganisms as well.

REFERENCES

Araújo, J. 1978. Anatomie comparée des systèmes glandulaires de défense chimique des Staphylinidae. *Archives de Biologie* 89:217–250.

Araújo, J., and J. M. Pasteels. 1987. Ultrastructure de la glande defensive d'*Eusphalerum minutum* Kraatz (Coleoptera: Staphylinidae). *Archives de Biologie Bruxelles* 98:15–34.

Happ, G. M., and C. M. Happ. 1973. Fine structure of the pygidial glands of *Bledius mandibularis* (Coleoptera: Staphylinidae). *Tissue and Cell* 5:215–231.

Jefson, M., J. Meinwald, S. Nowicki, K. Hicks, and T. Eisner. 1983. Chemical defense of a rove beetle *(Creophilus maxillosus)*. *Journal of Chemical Ecology* 9:159–180.

Jenkins, M. F. 1957. The morphology and anatomy of the pygidial glands of *Dianous coerulescens* Gyllenhal (Coleoptera: Staphylinidae). *Proceedings of the Royal Society London* 32:159–167.

Kellner, R. L. 2001. Suppression of pederin biosynthesis through antibiotic elimination of endosymbionts in *Paederus sabeus*. *Journal of Insect Physiology* 47:475–483.

———— 2002. Molecular identification of an endosymbiotic bacterium associated with pederin biosynthesis in *Paederus sabeus* (Coleoptera: Staphylinidae). *Insect Biochemistry Molecular Biology* 32:389–395.

Klinger, R., and U. Maschwitz. 1977. The defensive gland of Omaliinae (Coleoptera: Staphylinidae). I. Gross morphology of the gland and identification of the scent of *Eusphalerum longipenni* Erichson. *Journal of Chemical Ecology* 3:401–410.

Linsenmair, K. E., and R. Jander. 1963. Das Entspannungschwimmen bei *Velia* und *Stenus*. *Naturwissenschaften* 50:231.

Matsumoto, T., Y. M. Yanagiya, S. Maeno, and S. Yasuda. 1968. A revised structure of pederin. *Tetrahedron Letters* 60:6297–6300.

Pavan, M. 1963. Ricerche biologiche e mediche su pederina e su estratti purificati di *Paederus fuscipes* Curt. (Coleoptera Staphylinidae). Pavia: Industria Lito-Tipografiche M. Ponzio.

Piel, J. 2002. A polyketide synthase-peptide synthetase gene cluster from an uncultured bacterial symbiont of *Paederus* beetles. *Proceedings of the National Academy of Sciences USA* 99:14002–14007.

Piel, J., I. Höfer, and D. Hui. 2004. Evidence for a symbiosis island involved in horizontal acquisition of pederin biosynthetic capabilities by the bacterial symbiont of *Paederus fuscipes* beetles. *Journal of Bacteriology* 186:1280–1286.

Schildknecht, H., D. Krauss, J. Connert, H. Essenbreis, and N. Orfanides. 1975. The spreading alkaloid stenusin from the staphylinid *Stenus comma* (Coleoptera: Staphylinidae). *Angewandte Chemie—International Edition in English* 14:427.

Wheeler, J. W., G. M. Happ, J. Araújo, and J. M. Pasteels. 1972. Dodecalactone from rove beetles. *Tetrahedron Letters* 46:4635–4638.

40

Class **INSECTA**
Order **COLEOPTERA**
Family Cantharidae
Chauliognathus lecontei
A soldier beetle

Chauliognathus lecontei beetles mating.

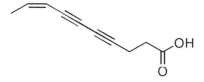

94. *(Z)*-Dihydromatricaria acid

The Cantharidae, known also as soldier beetles, or leather-wings, are soft-bodied, often gaudily colored insects, commonly found in aggregations on flowers. Well known are species of the genus *Chauliognathus*. Deliberate in their movements, and slow in flight, these beetles are easy to catch. Not surprisingly, they are chemically protected. Unlike many other beetles, which have a pair of defensive glands opening at the tip of the abdomen, *Chauliognathus* beetles have paired segmental glands, in the prothorax and in each of the first eight abdominal segments, with openings visible as small pores near the outer margins of the segments. When disturbed, they typically emit droplets of a white viscous fluid from the glands. Studies with two species, *C. lecontei* and *C. pennsylvanicus,* showed that these beetles are unacceptable to jays, grasshopper mice, carabid beetles, ants, and jumping spiders, but not consistently to preying mantids, assassin bugs, centipedes, or solpugids. Chemical analysis showed that the secretion in both these species contains *(Z)*-dihydromatricaria acid (**94**), an acetylenic compound.

A simple bioassay showed that *(Z)*-dihydromatricaria acid is deterrent to jumping spiders. Such spiders, after a period in captivity, can be fed by offering them individual fruit flies, freshly killed by freezing and suspended from a human hair. Ordinarily the spiders accept such flies, but they reject them if they are treated by addition of *(Z)*-dihydromatricaria acid. As little as 1 microgram of *(Z)*-dihydromatricaria acid, an equivalent of less than 2% of the amount of the acid present in a single *Chauliognathus,* can render a fly unacceptable. There can be little question, therefore, that *(Z)*-dihydromatricaria acid serves in defense.

Acetylenic compounds, including the methyl ester of dihydromatricaria acid, are known also from fungi and plants. Among the

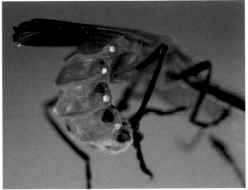

Left: A group of *Chauliognathus lecontei* beetles in Arizona, moving around the flowers of a stand of asclepiad plants on which they had aggregated. Right: A *C. lecontei* being held in forceps responds by emitting droplets of white defensive secretion from its abdominal glands.

plants are numerous Asteraceae, upon which species of *Chauliognathus* commonly aggregate. There remains some question, therefore, whether soldier beetles derive their *(Z)*-dihydromatricaria acid from plants or whether they synthesize the compound themselves. The presence of the compound in plants frequented by the beetles does not necessarily imply that the substance is obtained by the beetles from a botanical source. The beetles could themselves, some time in the remote past, have evolved the capacity to produce *(Z)*-dihydromatricaria acid. Having acquired this capacity could have made them tolerant of the compound and opened the way for their subsequent association with acetylenic acid–producing plants.

The *Cantharidae* are worldwide in distribution. There are some 5,000 species, in 135 genera. *Chauliognathus* is a widespread genus.

REFERENCES

Barth, R. 1958. Über die abdominalen Seitendrüsen von *Discodon cyanomelas* (Perty) (Coleoptera, Cantharidae, Silini). *Studia Entomologica* 1:487–495.

Brown, W. V., M. J. Lacey, and B. P. Moore. 1988. Dihydromatricariate-based triglycerides, glyceride ethers, and waxes in the Australian soldier beetle, *Chauliognathus lugubris* (Coleoptera: Cantharidae). *Journal of Chemical Ecology* 14:411–423.

Eisner, T., D. Hill, M. Goetz, S. Jain, D. Alsop, S. Camazine, and J. Meinwald. 1981. Antifeedant action of Z-dihydromatricaria acid from soldier beetles (*Chauliognathus* spp.). *Journal of Chemical Ecology* 7:1149–1158.

Meinwald, J., Y. C. Meinwald, A. M. Chalmers, and T. Eisner. 1968. Dihydromatricaria acid: acetylenic acid secreted by soldier beetle. *Science* 160:890–892.

Swain, T. 1963. *Chemical Plant Taxonomy.* New York: Academic Press.

Weatherston, J., and J. E. Percy. 1978. Venoms of Coleoptera. In S. Bettini, ed., *Arthropod Venoms.* Berlin: Springer-Verlag.

41

Class **INSECTA**

Order **COLEOPTERA**

Family Lampyridae

Photinus ignitus and *Photuris versicolor*

Fireflies

A firefly of the genus *Photuris*. The light organ is denoted by the ivory-colored rear of the abdomen.

Fireflies are not quite what their name implies. They are not flies but beetles, and although they emit light as fires do, their light is cold rather than hot. There are some 2,000 firefly species worldwide, belonging to about 100 genera.

Fireflies depend on their light organs for courtship. Individuals that fly about in the night are for the most part males, and the flashes they emit are meant to advertise their presence and availability to the females.

Photinus ignitus (left) and *Photuris versicolor* (right).

95. Lucibufagin C

96. *N*-Methylquinolinium-2-carboxylate

Several species of firefly may coexist in the same area. Confusion is prevented because each species has its own communicative code. Males emit species-specific flash sequences—combinations of short and long light pulses—by which the females recognize their potential mates. The females in turn respond by emitting single flashes, which the males can tell apart because females of different species differ in the timing of their responses—in the delay with which they emit their own flash following the end of the male's pulse sequence.

Females, when on the alert for males, are ordinarily poised on vegetation. When a male receives an appropriately timed response to his flashes, he flies toward the female and mating typically follows.

Fireflies are chemically protected. Species of the genus *Photinus,* including *P. ignitus* from the eastern United States, have been shown to contain certain distasteful steroids called lucibufagins, for example, lucibufagin C (**95**). The compounds are present in both sexes of *Photinus,* and provide the fireflies with protection against spiders, birds, and probably other predators.

The presence of lucibufagins in one genus of firefly prompted a search for the compounds in other genera, and what was found was peculiar. Lucibufagins were also present in a firefly of the genus *Photuris* (specifically in *P. versicolor*), but in substantial quantity in the female alone and, oddly, only after the mating season was well under way. There was a reason for this. *Photuris* fireflies are known as firefly femmes fatales. They attract firefly males of two kinds—their own kind for mating purposes, and those of the genus *Photinus* for food. They attract *Photuris* males by adhering to their own intraspecific code—by timing their flash the *Photuris* way—but they draw *Photinus* males by resorting to the signal delay characteristic of *Photinus.* Male *Photinus* are deceived by the fakery and fly toward the *Photuris* females, only to be overpowered and eaten. It was long thought that the only benefit derived by the *Photuris* female from this practice was nutritional, but it is now known that by feasting on *Photinus* the female also acquires her victims' lucibufagins. The female is able to use these chemicals to advantage. Against jumping spiders, for example, she is quite helpless, unless she has previously "dined" on *Photinus.* The acquired lucibufagins become her defense.

Female *Photuris* fireflies can feed on more than one *Photinus* male in the course of their lives. The more males they eat, the more lucibufagin they acquire. Females share the acquired chemical with their offspring: they transmit lucibufagins to the eggs, which are protected against ants as a result.

Female *Photuris* fireflies emerge from the pupa lucibufagin-free,

A *Photuris versicolor* female devouring a *Photinus ignitus* male.

which explains why it is not until later in the season, after they have had the opportunity to catch some *Photinus,* that field-collected *Photuris* females contain lucibufagins. *Photuris* females are not entirely defenseless if they fail to catch *Photinus.* The females also produce a defensive compound of their own, a quinoline derivative (**96**), which probably provides them with some measure of protection and which they also share with their eggs. *Photuris* males too produce this compound, which may be of importance to them, since these males do not, as a matter of routine, acquire lucibufagins by eating other males.

REFERENCES

Eisner, T., D. F. Wiemer, L. W. Haynes, and J. Meinwald. 1978. Lucibufagins: defensive steroids from the fireflies *Photinus ignitus* and *P. marginellus* (Coleoptera: Lampyridae). *Proceedings of the National Academy of Sciences USA* 75:905–908.

Eisner, T., M. A. Goetz, D. E. Hill, S. R. Smedley, and J. Meinwald. 1997. Firefly "femmes fatales" acquire defensive steroids (lucibufagins) from their firefly prey. *Proceedings of the National Academy of Sciences USA* 94:9723–9728.

Eisner, T., C. Rossini, A. González, V. K. Iyengar, M. V. S. Siegler, and S. R. Smedley. 2002. Paternal investment in egg defense. In M. Hilker and T. Meiners, eds., *Chemoecology of Insect Eggs and Egg Deposition.* Berlin: Blackwell Publishing.

González, A., J. F. Hare, and T. Eisner. 2000. Chemical egg defense in *Photuris* firefly "femmes fatales." *Chemoecology* 9:177–185.

González, A., F. Schroeder, J. Meinwald, and T. Eisner. 1999. Nmethylquinolinium 2-carboxylate, a defensive betaine from *Photuris versicolor* fireflies. *Journal of Natural Products* 62:378–380.

Lloyd, J. E. 1965. Aggressive mimicry in *Photuris:* firefly femmes fatales. *Science* 149:653–654.

——— 1975. Aggressive mimicry in *Photuris* fireflies: signal repertoires by femmes fatales. Science 187:452–453.

42

Class INSECTA
Order COLEOPTERA
Family Lycidae
Calopteron reticulatum
The banded net-winged beetle

Calopteron reticulatum.

A *Calopteron reticulatum* bleeding from a wing injury.

The members of the family Lycidae make up an interesting group of beetles. Worldwide in distribution and characteristically soft-bodied, they are usually black with red or orange markings, and as a result are quite conspicuous. Slow in flight and generally lethargic, they tend to occur in moist habitats, most commonly on leaves or flowers, and not infrequently in aggregations. The term "net-winged" as applied to lycids refers to the network of prominent longitudinal and less prominent transverse ridges characteristic of the elytra of these beetles. Lycids are believed to feed primarily on pollen and nectar, and are relatively short-lived as adults. The larvae occur under bark and in rotten wood, where they may feed on slime molds and fungal material. The family includes about 150 genera and 3,500 species.

Lycids have long been known to figure as models in mimicry complexes. This fact, coupled with the beetles' showiness and lethargy, led entomologists early on to suggest that lycids were chemically protected, as they are now known to be. Lycids are mimicked

97. 2-Methoxy-3-isopropylpyrazine

98. Lycidic acid

99. 1-Methyl-2-quinolone   100. 3-Phenylpropanamide

by beetles of several families, including the Cerambycidae, Cleridae, Chrysomelidae, Meloidae, Buprestidae, Lampyridae, and Belidae, as well as by diverse lepidopterans. Some of these mimics, including the meloids and lampyrids, may themselves be distasteful.

Lycids derive their noxiousness from a diversity of chemical factors. *Calopteron reticulatum,* for example, a species occurring in North America east of the Rockies, produces substances called pyrazines (for example, **97**) that appear to account for the faint but distinct stink of the beetle. The animal also produces a series of unusual fatty acids (for example, lycidic acid; **98**) present throughout its body, which may be responsible for its distastefulness. Lycids may not all produce the same defensive compounds. The Australian lycid *Metriorrhynchus rhipidius* produces a quinolone derivative (**99**) and a phenylamide (**100**) not found in any of the North American lycids that have been studied. Lycids have no defensive glands, but they tend to bleed when seized, usually first from the wings,

Lycus fernandezi (top left), and its cerambycid mimic, *Elytroleptus apicalis* (top right). Bottom: *E. apicalis* feeding on a *Lycus*.

the ridges of which are hollow and subject to rupture. It is by bleeding that lycids externalize their unpalatable factors. Odorous agents such as the pyrazines are apparently emitted on a continuous basis.

A remarkable relationship exists in the southwestern United States between certain lycids and their mimics. The lycids, all members of the genus *Lycus,* form dense aggregations during the summer months on various flowers, where they are joined by their mimics. These include moths, as well as certain long-horned beetles (Cerambycidae) of the genus *Elytroleptus.* What is interesting is that *Elytroleptus* feed on the model lycids. Such behavior could potentially put *Elytroleptus* at risk, since with reduced numbers of model lycids in their surroundings, the cerambycids might not derive benefits from their mimetic appearance. What makes this unsual relationship possible is that the *Elytroleptus* occur in very small numbers relative to the model lycids, with the result that they never inflict a heavy numerical toll upon the lycids. *Elytroleptus* beetles, unlike the model lycids, are palatable to predators. Contrary to what you might expect, *Elytroleptus* do not become distasteful as a consequence of eating lycids. They are just as acceptable to spiders after consuming a lycid as they are after a prolonged period of fasting.

REFERENCES

Darlington P. J. 1938. Experiments on mimicry in Cuba, with suggestions for future study. *Transactions of the Royal Entomological Society London* 87:681–695.

Eisner, T., and F. C. Kafatos. 1962. Defense mechanisms of arthropods. X. A pheromone promoting aggregation in an aposematic distasteful insect. *Psyche* 69:53–61.

Eisner, T., F. C. Kafatos, and E. G. Linsley. 1962. Lycid predation by mimetic adult Cerambycidae (Coleoptera). *Evolution* 16:316–324.

Eisner, T., N. F. R. Snyder, F. Schroeder, J. Grant. Unpublished data on *Lycus* and *Elytroleptus.*

Linsley, E. G., T. Eisner, and A. B. Klots. 1961. Mimetic assemblages of sibling species of lycid beetles. *Evolution* 15:15–29.

Moore, B. P., and W. V. Brown. 1981. Identification of warning odour components, bitter principles, and antifeedants in an aposematic beetle: *Metriorrhynchus rhipidius* (Coleoptera: Lycidae). *Insect Biochemistry* 11:493–499.

———— 1989. Graded levels of chemical defence in mimics of lycid beetles of the genus *Metriorrhynchus* (Coleoptera). *Journal of the Australian Entomological Society* 28:229–233.

43

Class **INSECTA**

Order **COLEOPTERA**

Family Elateridae

Alaus myops

The eyed elater

An unidentified elaterid beetle.

Elaterids are familiar to naturalists because of their peculiar habit of hurling themselves into the air when placed belly side up on a solid surface. They resort to the behavior in order to regain their upright stance, and usually succeed in doing so. Jumping occurs with an accompanying clicking sound; hence the name "click beetles" applied to elaterids.

The body of elaterids is hinged at the juncture of the prothorax and mesothorax. This makes it possible for a click beetle to flex its front end ventrally, which is what it does, abruptly, when executing

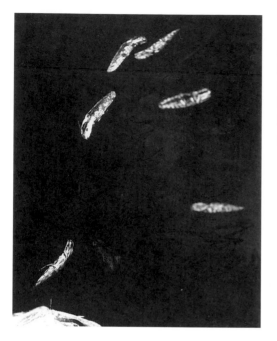

An elaterid beetle *(Alaus myops),* weighing 1.1 grams, flipping itself into the air (to a height of 20 centimeters) after being placed belly side up on a solid surface. The photograph was taken by repeated electronic flash exposures.

the jump. Every elaterid has a conspicuous ventral peg that projects posteriorly from the prosternum, and fits snugly into an elongate pit of the mesosternum. Ordinarily, when the beetle's body is straightened out in its normal posture, the tip of the peg is held by a "catch" at the anterior end of the pit. To trigger the jump, the beetle tenses a special set of muscles until enough force is built up to release the peg from its catch. When this occurs, the peg slides into the depth of the pit, in one quick motion that also causes the beetle's front end to flex ventrally. The flexion causes the beetle's back to smack against the ground so that the beetle is hurled into the air. The audible click, engendered when the peg strikes the depth of the pit, attests to the very considerable force that is released by the beetle's flexion.

The primary function of clicking in elaterids is probably defensive. Although the beetles may certainly rely on the behavior to regain their upright stance, they probably make use of it more often to startle predators, or to flip themselves free from a predator's hold. Experiments in which elaterids of various sizes, including specimens of the very large *Alaus myops,* were offered to orb-weaving spiders

A photographic double exposure depicting the motion executed by an elaterid when "clicking." The beetle's front end (head plus prothorax) ordinarily projects straight forward (the upper position of the front end in the photograph). When clicking, the beetle abruptly flexes the front end ventrally, as shown.

were revealing. The spiders invariably attacked, but the beetles commenced clicking the moment they were seized, and in many cases succeeded in freeing themselves. Experiments with wolf spiders gave similar results, as did tests with a mouse and a jay, although with these two vertebrates the elaterids escaped only rarely.

The jolt effected by elaterids when they click is impressive. It certainly has a startling effect on humans, and it could similarly affect predators. But with smaller predators, the beetles may literally kick themselves free. The height to which elaterids project themselves when flipping provides an indication of the force that is unleashed when the beetles click.

Elaterids occur in all parts of the world and are often locally abundant. Some 400 genera and 9,000 species are known. The larvae of some species are serious agricultural pests.

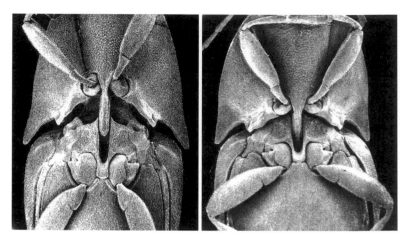

A ventral view of an elaterid showing the prosternal peg, both in its ordinary "hold" position (left) and after it has been driven into the pit in the course of a "click" (right).

REFERENCES

Eisner, T., unpublished predation tests with elaterid beetles.
Evans, M. E. G. 1972. The jump of the click beetle (Coleoptera, Elateridae)—a preliminary study. *Journal of Zoology* (London) 167:319–336.
——— 1973. The jump of the click beetle (Coleoptera, Elateridae)—energetics and mechanics. *Journal of Zoology* (London) 169:181–194.

44

Class **INSECTA**

Order **COLEOPTERA**

Family Buprestidae

Acmaeodera pulchella

The flat-headed baldcypress sapwood borer

Acmaeodera amplicollis.

The sight of flower-visiting insects with translucent membranous wings and an abdomen bearing transverse yellow or milky-white stripes conjures up the thought of bees and wasps and of the sting such insects can inflict. "Be cautious," is the message we derive from such warnings, which are likely to be heeded by predators as ↳ well. But there are insects that share the visual characteristics of these Hymenoptera without also having a sting apparatus, insects that are hymenopteran mimics and for that reason protected from predation. Examples include flies from a number of families (for instance, the Syrphidae, Asilidae, and Bombiliidae), which have brightly striped abdomens and consequently a distinct wasp-like appearance, particularly when flying.

There exist beetles, too, that are hymenopteran mimics. Interesting among these are members of the genus *Acmaeodera,* of the family Buprestidae. Buprestids are known as metallic wood-boring beetles, in reference to the bright shine that gives so many of them a jewel-like appearance. Not surprisingly, buprestids are a favorite of beetle collectors.

Buprestids, for the most part, fly as beetles usually do, with their solid front wings (elytra) held out to the sides to provide lift and their hindwings vibrating to provide thrust. *Acmaeodera* are exceptional in that they fly with the hindwings only. Their elytra are fused along the dorsal midline, and are held as a shield over the abdomen even when the beetles are in flight. When airborne, therefore, with only their hindwings in service, *Acmaeodera* have a distinct hymenopteran appearance. But what is most misleading about their appearance is that the elytra bear the banded color patterns characteristic of the hymenopteran abdomen. Some *Acmaeodera,* such as *A. pulchella,* have yellow stripes on the elytra, while others have stripes that are almost white. Their wasp-like appearance, when on the wing, is uncanny.

Little is known about the palatability of *Acmaeodora,* but such evidence as exists suggests that the beetles are not intrinsically distasteful. They are therefore Batesian rather than Müllerian mimics, in the sense that they are imitative of an inedible insect without themselves being noxious.

Flying with the hindwings alone has required some specialization on the part of *Acmaeodera.* To enable the hindwings to exercise their beat, notches are provided on the margins of the elytra through which the hindwings project laterally when the beetles are in flight.

Buprestids include some 15,000 species in 400 genera worldwide. The North American *Acmaeodera* constitute some 150 species, mostly from the Southwest.

Also flying with the hindwings only, and seemingly imitative of

Top: *Acmaeodera pulchella* in tethered flight. The elytra, bearing the transverse yellow stripes that adorn many a hymenopteran abdomen, are permanently locked to form a shield over the beetle's abdomen. Bottom: A wasp *(Dolichovespula arenaria)* in tethered flight, its yellow abdominal stripes clearly on display.

bees, are scarab beetles of the tribe Cetoniini. One species of this group, *Euphoria limbalis,* has the same darting and hovering flight as a large carpenter bee *(Xilocopa micans)* that shares its habitat. Cetoniines too have notched elytra to provide for the lateral projection of the hindwings during flight.

REFERENCES

Silberglied, R. E., and T. Eisner. 1969. Mimicry of Hymenoptera by beetles with unconventional flight. *Science* 163:486–488.

45

Class INSECTA
Order COLEOPTERA
Family Coccinellidae
Cycloneda sanguinea
A ladybird beetle

The ladybird beetle *Cycloneda sanguinea*.

There are two especially vulnerable stages in the life cycle of an insect—the egg and the pupa. In both, the insect is anchored and unable to escape from impending danger, with the result that it must confront hazards directly. Not surprisingly, eggs and pupae are often

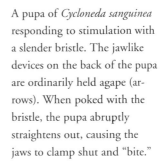

A pupa of *Cycloneda sanguinea* responding to stimulation with a slender bristle. The jawlike devices on the back of the pupa are ordinarily held agape (arrows). When poked with the bristle, the pupa abruptly straightens out, causing the jaws to clamp shut and "bite."

protected, but in ways that can vary considerably in insects of different groups. Eggs may derive protection from noxious chemicals provided by the parents (see, for instance, Chapters 23 and 62), while pupae may achieve safety by being camouflaged, encased within cocoons, or hidden in earthen chambers. Most remarkable, perhaps, are certain insect pupae that have the ability to "bite." This is not to say that these pupae have operative mouthparts. Rather, they possess the functional equivalents of jaws, pinching devices that they can set in motion when assaulted. The devices have been called gin traps because of their similarity to the contraptions once used extensively to snare game. Insect gin traps take the form of clefts with sharp, hardened borders, capable of being clamped shut and inflicting bites. They are usually present in the pupal abdomen and have evolved independently, for the same obvious purpose, in a number of beetles and moths.

Typical gin traps are found in the pupae of ladybird beetles, as for example in *Cycloneda sanguinea,* a species ranging broadly throughout the United States and into South America. The traps take the form of four deep clefts, situated dorsally on the abdomen between the segments. The anterior margin of each cleft is densely beset with tiny teeth, while the posterior margin is smooth and sharp. When the pupa is at rest, with the body flexed and recumbent against the substrate, the traps are agape and "set" for action. Stimulating such a pupa with a brush will cause it to straighten out abruptly, in a quick flipping action that causes the clefts to clamp shut. Even mild stimulation will trigger the response, as anyone can verify by "tickling" a pupa with a human hair. The pupa will almost certainly respond no matter where it is tickled, but it is guaranteed to "bite" if the hair is inserted directly into one of the clefts. The experiment is easily performed. Ladybird beetle pupae are not hard to find. They are most commonly located near the colonies of aphids, upon which larval and adult ladybird beetles frequently prey.

Experiments in which *C. sanguinea* pupae were exposed to ants showed dramatically that the traps are effective. The pupae responded quickly when touched by an ant, and they usually did so by making several flips in quick succession. The ants were deterred by the flips, particularly when their appendages were caught in the traps and pinched.

Gin traps occur also in the pupae of desert-dwelling tenebrionid beetles of the genus *Blaps* and in pupae of *Manduca sexta,* a sphinx moth frequently maintained in culture by biologists. In *Blaps,* the gin traps take the form of intersegmental pinching devices that are activated by lateral flexions of the abdomen. In *M. sexta,* they are fashioned as sharp-margined, abdominal slits that are also clamped shut when the pupa is disturbed.

An antenna of an attacking ant caught in a gin trap of *Ccycloneda sanguinea.*

A pupa of a European tenebrionid beetle of the genus *Blaps* responding to being poked with a wooden probe by clamping down with its sharp-toothed abdominal gin traps.

REFERENCES

Bate, C. M. 1973. The mechanism of the pupal gin trap. I. Segmental gradients and the connections of the triggering sensilla. *Journal of Experimental Biology* 59:95–107.

———— 1973. The mechanism of the pupal gin trap. II. The closure movement. *Journal of Experimental Biology* 59:109–119.

———— 1973. The mechanism of the pupal gin trap. III. Interneurons and the origin of the closure mechanism. *Journal of Experimental Biology* 59:121–135.

Eisner, T., and M. Eisner. 1992. Operation and defensive role of "gin traps" in a coccinellid pupa *(Cycloneda sanguinea)*. *Psyche* 99:265–273.

Hinton, H. E. 1946. The "gin traps" of some beetle pupae: a protective device that appears to be unknown. *Transactions of the Royal Entomological Association London* 97:473–496.

Waldrop, B., and. R. B. Levine. 1989. Development of the gin trap reflex in *Manduca sexta:* a comparison of larval and pupal motor responses. *Journal of Comparative Physiology* 165:743–754.

46

Class **INSECTA**

Order **COLEOPTERA**

Family Coccinellidae

Epilachna varivestis

The Mexican bean beetle

Epilachna varivestis.

The family Coccinellidae includes the ladybird beetles, perhaps the most beneficial of all Coleoptera. They are for the most part predaceous, both as larvae and adults, and play a major role in controlling the populations of aphids and scale insects, many of which are pests. Some coccinellids have been used in agricultural control programs.

Coccinellids include also some species that are herbivorous, and these may themselves be pests. A notable example is the Mexican bean beetle, *Epilachna varivestis,* a common pest of bean plants. The

Epilachna varivestis, tethered, in ventral view. The beetle is responding to the pinching of the right and left hindlegs by emitting droplets of blood from the knee joints of the legs stimulated.

Mexican bean beetle is a hardy insect, well defended against predation at all stages of development. It has been the subject of intensive study, and its defenses, which are for the most part chemical, are well understood.

E. varivestis is protected primarily by alkaloids. One such compound, 1-(2-hydroxyethyl)-2-(12-aminotridecyl)-pyrrolidine (**101**) occurs in the eggs and larvae, and is retained systemically into the adult stage. But the compound does not persist in the adult; it is replaced, beginning about the second day after emergence from the pupa, by a homotropane alkaloid, euphococcinine (**102**). In its immature stages the beetle also contains a number of lower-order homologues of **101**, and as an adult, a series of piperidine derivatives (for example, **103**).

The adult Mexican bean beetle, which contains defensive alkaloids in the blood, reflex-bleeds from the knee joint when disturbed. It is programmed to do so in controlled fashion. If one leg of the beetle is pinched with forceps, the beetle bleeds from that leg only. If a number of legs are pinched, it bleeds from these, in the sequence in which they are pinched. The larvae do not reflex-bleed, but their

101. 1-(2-Hydroxyethyl)-2-(12-aminotridecyl)-pyrrolidine   **102.** Euphococcinine

103. 6-Methyl-2-(2-oxopropyl)-piperidine   **104.** Pseudopelletierine

bodies are covered with hollow branched spines that rupture readily when the larvae are attacked, with consequent emission of blood.

Data are lacking on the extent to which the various alkaloids of *E. varivestis* contribute to the beetle's defense. But one of the compounds, euphococcinine (**102**), was demonstrated to have considerable defensive potential. When it was added to a sugar solution, the compound rendered the solution unacceptable to ants, and when added to freshly killed fruit flies offered as bait to jumping spiders, it made the bait unacceptable to the spiders. Although no tests have been done with vertebrates, the compound may well be deterrent to these animals as well. An alkaloid closely related to euphococcinine, pseudopelletierine (**104**), from pomegranate bark, is extremely bitter to humans.

E. varivestis lacks the pupal "gin traps" (see Chapter 45) present in other beetles of the family Coccinellidae. Instead, its pupae are chemically protected. If an *E. varivestis* pupa is viewed at low magnification with a stereomicroscope and illuminated so that it is partly back-lit, the observer will note that its surface is covered with tiny hairs, bearing droplets of an oily fluid at the tip. This fluid, which can be taken up cleanly into glass capillary tubes, has been analyzed

A pupa of *Epilachna varivestis* in dorsal view, showing the dense covering of glandular hairs.

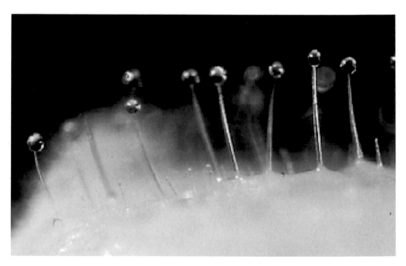

Close-up view of the glandular hairs of an *Epilachna varivestis* pupa.

and shown to contain a series of closely related alkaloids, known from nowhere else in nature, for which the name azamacrolides was coined. The principal azamacrolide in the fluid is epilachnene (**105**), which in assays with insects was found to be repellent and irritating.

The pupal hair secretion varies in interesting ways from species to species. In a close relative of *E. varivestis,* the squash beetle, *Epilachna borealis,* the fluid contains a mixture of several hundred macrocyclic polyamines (the polyazamacrolides, or PAMLs), compounds of up to enormous ring size (**106–108**), which seem to

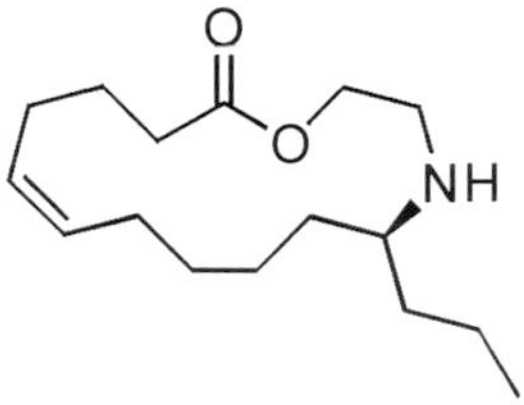

105. Epilachnene

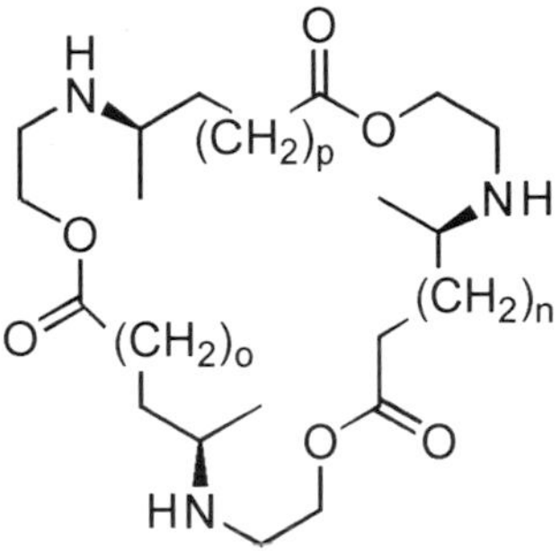

106. A Trimeric polyazamacrolide

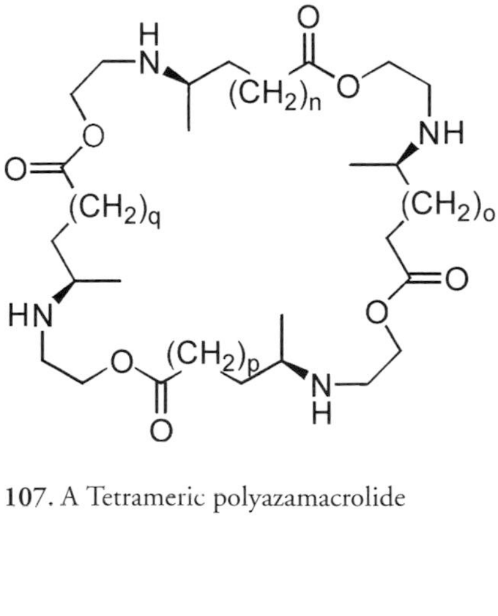

107. A Tetrameric polyazamacrolide

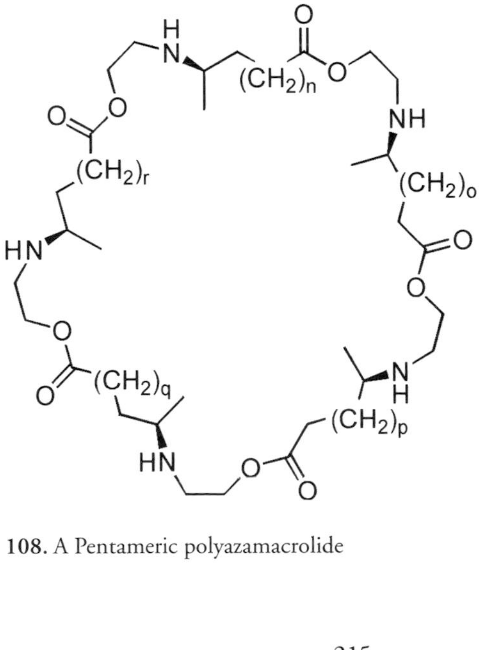

108. A Pentameric polyazamacrolide

215

be generated in the secretion by random assembly of three simple molecular building blocks: (ω-1)-hydroxyethylamino-alkanoic acids (**109–111**).

In a third coccinellid beetle, *Subcoccinella 24-punctata,* a close relative of the *Epilachna* species, the pupal hairs secrete yet other PAMLs (for example, **112, 113**), presumably also by random combinatorial synthesis, but from a different set of building blocks (**114, 115**).

From a chemical point of view, coccinellids have proved to be of considerable interest. The beetles were long known to be bitter in taste, and suspected to be shunned by predators, so it came as no surprise when recent work revealed them to be the repository of a great diversity of alkaloids. Interestingly, coccinelline (**116**), one of the first two alkaloids characterized from coccinellid beetles, was isolated by testing the fractions obtained in the course of the purification procedure for both deterrency to ants and bitterness to humans. Although only a relatively few alkaloids from coccinellids have been tested for defensive efficacy, evidence is mounting that these compounds are broadly protective against both insectan and vertebrate predators. Ants are probably among the major threat to coccinellids at all life stages, and ants have consistently proven to be deterred when exposed to coccinellid alkaloids. Epilachnine (**105**), for instance, and coccinelline (**116**) are potently active against ants. Coccinelline, moreover, is active also against birds.

Coccinellids can themselves produce much of their defensive arsenal. Thus, as has been demonstrated by the feeding of radio-labeled precursors, both coccinelline and azamacrolides are produced in their respective coccinellid sources by endogenous synthesis. But in some instances, coccinellids make defensive use of substances they incorporate from exogenous sources. This has been demonstrated for *Hyperaspis trifurcata* (see Chapter 29), which derives carminic acid (**60**) from the cochineal bugs it eats, and for certain other coccinellids that obtain selected alkaloidal or steroidal toxins by eating aphids that derive these compounds from plants.

An intriguing possibility is that some of the alkaloids of coccinellids are put to secondary use by dendrobatid frogs that feed on the beetles. Dendrobatid frogs include species that store toxins in

109. 8-(2-Hydroxyethylamino)-nonanoic acid

110. 9-(2-Hydroxyethylamino)-decanoic acid

111. 10-(2-Hydroxyethylamino)-undecanoic acid

112. A Tetra-unsaturated polyazamacrolide

113. A Tri-unsaturated polyazamacrolide

114. An (2-Hydroxyethylamino)-alkadienoic acid

115. An (2-Hydroxyethylamino)-alkenoic acid

116. Coccinelline

the skin, and there is increasing evidence that such compounds are derived by the frogs from arthropods they eat (see Chapter 68).

With about 500 genera and 4,500 species, the Coccinellidae make up a large family.

REFERENCES

Attygalle, A. B., K. D. McCormick, C. L. Blankespoor, T. Eisner, and J. Meinwald. 1993. Azamacrolides: a family of alkaloids from the pupal defensive secretion of a ladybird beetle (*Epilachna varivestis*). *Proceedings of the National Academy of Sciences USA* 90:5204–5208.

Attygalle, A. B., S.-C. Xu, K. D. McCormick, J. Meinwald, C. L. Blankespoor, and T. Eisner. 1993. Alkaloids of the Mexican bean beetle *Epilachna varivestis* (Coccinellidae). *Tetrahedron* 40:9333–9342.

Attygalle, A. B., A. Svatos, M. Veith, J. J. Farmer, J. Meinwald, S. Smedley, A. González, and T. Eisner. 1999. Biosynthesis of epilachnene, a macrocyclic defensive alkaloid of the Mexican bean beetle. *Tetrahedron Letters* 55:955–966.

Daly, J. W. 1995. The chemistry of poisons in amphibian skin. In T. Eisner and J. Meinwald, eds., *Chemical Ecology.* Washington, D.C.: National Academy Press.

——— 1998. Thirty years of discovering arthropod alkaloids in amphibian skins. *Journal of Natural Products* 61:162–172.

Daly, J. W., H. M. Garraffo, T. F. Spande, C. Jaramillo, and A. S. Rand. 1994. Dietary source for skin alkaloids of poison frogs (Dendrobatidae)? *Journal of Chemical Ecology* 20:943–955.

Daly, J. W., S. I. Secunda, H. M. Garraffo, T. F. Spande, A. Wisnieski, and J. F. Cover, Jr. 1994. An uptake system for dietary alkaloids in poison frogs (Dendrobatidae). *Toxicon* 32:657–663.

Eisner, T., M. Goetz, D. Aneshansley, G. Ferstandig-Arnold, and J. Meinwald. 1986. Defensive alkaloid in blood of Mexican bean beetle *(Epilachna varivestis). Experientia* 42:204–207.

King, A. G., and J. Meinwald. 1996. Review of the defensive chemistry of coccinellids. *Chemical Reviews* 96:1105–1122.

Proksch, P., L. Witte, V. Wray, and T. Hartmann. 1993. Ontogenetic variation of defensive alkaloids in the Mexican bean beetle *Epilachna varivestis* (Coleoptera: Coccinellidae). *Entomologia Generalis* 18:1–7.

Rossini, C., A. González, J. Farmer, J. Meinwald, and T. Eisner. 2000.

Anti-insectan activity of epilachnene, a defensive alkaloid from the pupae of Mexican bean beetles *(Epilachna varivestis)*. *Journal of Chemical Ecology* 26:391–397.

Schröder, F., J. J. Farmer, A. B. Attygalle, J. Meinwald, S. R. Smedley, and T. Eisner. 1998. Combinatorial chemistry in insects: a library of defensive macrocyclic polyamines. *Science* 281:428–431.

Schröder, F. C., J. J. Farmer, S. R. Smedley, T. Eisner, and J. Meinwald. 1998. Absolute configuration of the polyazamacrolides, macrocyclic polyamines produced by a ladybird beetle. *Tetrahedron Letters* 39:6625–6628.

Schroeder, F. C., S. R. Smedley, L. K. Gibbons, J. J. Farmer, A. B. Attygalle, T. Eisner, and J. Meinwald. 1998. Polyazamacrolides from ladybird beetles: ring-size selective oligomerization. *Proceedings of the National Academy of Sciences USA* 95:13387–13391.

Tursch, B., D. Daloze, M. Dupont, J. M. Pasteels, and M. C. Tricot. 1971. A defense alkaloid in a carnivorous beetle. *Experientia* 27:1380–1381.

47

Class INSECTA
Order COLEOPTERA
Family Meloidae
Epicauta (an unidentified species)
A blister beetle

An aggregation of blister beetles (an unidentified *Epicauta* species) on a solanaceous plant in Arizona.

The beetles of the family Meloidae, the blister beetles, are an interesting lot. Known since antiquity for the toxic substance they contain, the infamous chemical cantharidin, or "Spanishfly" (117), meloids have been the subject of countless anecdotes, mostly revolving around the effect of Spanishfly on humans. Cantharidin has long been purported to be an aphrodisiac. Taken orally by men, the substance causes erections, but a Viagra it is not. Quite to the contrary, the compound induces severe irreversible damage to the renal and reproductive systems, and is lethal (to both men and women) in very small quantities (100 milligrams of cantharidin, the equivalent of what can be extracted from a few meloids, is said to be fatal to humans). Cantharidin is also a potent blistering agent. The compound still finds some use in dermatological medicine, but claims that it has general curative properties appear not to be substantiated.

Cantharidin serves meloids as a defensive agent. Individual meloids may contain as much as 17 milligrams of cantharidin, or 10% of live weight. The compound is stored in the blood and reproductive organs of the beetles, and the beetles commonly reflex-bleed (see also Chapter 46) when disturbed. If seized by a leg, they emit a droplet of blood from the knee joint of that leg only. Held by the body instead, they may bleed from all legs at once, as well as sometimes from the neck and elsewhere.

Meloid blood is deterrent to predators, and appears to derive its activity mostly from its contained cantharidin. Both spiders and ants have been shown to reject food that is laced with cantharidin at concentrations far below those at which the compound occurs in meloid blood.

To judge from biochemical studies, it seems that cantharidin is synthesized by the males only, at least in some meloids. The females

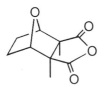

117. Cantharidin

A blister beetle (an unidentified *Epicauta* species) responding to being grasped with forceps by reflex-bleeding from the knee joints of several legs.

receive cantharidin from the males with the sperm package at mating, in amounts sufficient to protect both themselves and their eggs. Indeed, if radioactive molecular precursors of cantharidin are injected into male meloids, radioactive cantharidin is later detected not just in these males, but in their female partners and in the eggs they lay.

Considering the toxicity of cantharidin, it is remarkable that there should even exist predators able to feed on meloids with impunity. Predators tolerant of cantharidin include certain spiders, frogs, quail, and the European hedgehog. Why these animals should be able to tolerate cantharidin remains unexplained.

Even more remarkable is the existence of insects that are attracted to cantharidin. Such insects, which can be lured to cantharidin-baited traps, feed on cantharidin, and derive protection as a result (see Chapter 48).

There are some 3,000 species of meloids worldwide, belonging to about 120 genera.

REFERENCES

Blodgett, S. L., J. E. Carrel, and R. A. Higgins. 1991. Cantharidin content of blister beetles (Coleoptera: Meloidae) collected from Kansas alfalfa and implications for inducing cantharidiasis. *Environmental Entomology* 20:776–780.

Carrel, J. E., and T. Eisner. 1974. Cantharidin: potent feeding deterrent to insects. *Science* 183:755–757.

Eisner, T., J. Conner, J. E. Carrel, J. P. McCormick, A. J. Slagle, C. Gans, and J. C. O'Reilly. 1990. Systemic retention of ingested cantharidin by frogs. *Chemoecology* 1:57–62.

Smedley, S. R., C. L. Blankespoor, Y. Yang, J. E. Carrel, and T. Eisner. 1996. Predatory response of spiders to blister beetles (family Meloidae). *Zoology* 99:211–217.

McCormick, J. P., and J. E. Carrel. 1987. Cantharidin biosynthesis and function in meloid beetles. In G. D. Prestwich and G. J. Blomquist, eds., *Pheromone Biochemistry.* New York: Academic Press.

48

Class INSECTA
Order COLEOPTERA
Family Pyrochroidae
Neopyrochroa flabellata
A fire-colored beetle

The head of a *Neopyrochroa flabellata* male, showing the glandular cleft.

N*eopyrochroa flabellata* is a member of the Pyrochroidae, the fire-colored beetles. The family is relatively small (about 100 species, 10 genera) and *N. flabellata* is one of its larger representatives. The beetle is also a cantharidiphile, meaning that it is attracted to cantharidin (117), the defensive chemical known as Spanishfly, produced by blister beetles (see Chapter 47). Cantharidiphiles include

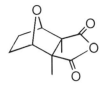

117. Cantharidin

also beetles from other families (for instance, the Anthicidae), as well as insects from other orders (Hemiptera, Diptera, and Hymenoptera). Recent work has shown that cantharidiphiles actually ingest cantharidin and that they put the compound to defensive and pheromonal use. The details of this chemical utilization have been worked out in a number of species, including *N. flabellata.*

N. flabellata beetles are readily caught in traps baited with cantharidin. Interestingly, it is only the males that are drawn to cantharidin, which they avidly ingest. Males that eat cantharidin secrete some of the chemical into a deep cleft, situated on the front of the head between the eyes. They retain the remainder of the compound internally, storing it within the accessory glands of the reproductive system.

The acquired cantharidin gives the male pyrochroid an advantage in courtship. When a female *N. flabellata* encounters a male, she promptly subjects him to a "cantharidin test." She grasps the male by his front end, inserts her mandibles into the cephalic cleft, and then, as she holds him in her grip, proceeds to eat the entire contents of the cleft. If the cleft contained cantharidin, the female becomes docile and yields to the male's copulatory attempts. But if the male is cantharidin-free, which the female can sense by inspecting the male's cleft, she refuses him, no matter how persistently he tries to mount her.

The cantharidin in the cleft provides the male with the means of "telling" the female that he is cantharidin-laden. The chemical serves as an enticing agent, a promissory note as it were, by which the female is informed that the male has additional cantharidin in store for her, but only if she accepts him as a mate. Indeed, at mating, the male transmits to the female, with the sperm package, the

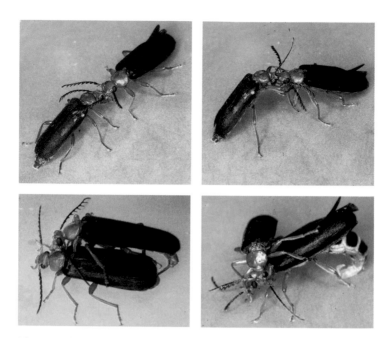

The courtship ritual of *Neopyrochroa flabellata*. Top: Male and female make head-to-head contact, following which the female feeds from the male's cephalic cleft. Bottom: The female subsequently accepts the male if she has detected cantharidin in his cephalic cleft, or rejects him if she found the cleft to be cantharidin-free.

full load of cantharidin he had stored away in the accessory glands. His strategy is evidently to withhold bestowing the bulk of his chemical gift until he is assured of acceptance of his sperm.

Much of the cantharidin that the female receives from the male at mating is transmitted by her to the eggs. Cantharidin is a potent feeding deterrent to ants and carabid beetles, as well as probably to a diversity of other predators, vertebrates included. There can be little doubt, therefore, that the *N. flabellata* eggs benefit from being cantharidin-endowed. Does a male have assurance that his cantharidin gift will be allocated to eggs of his siring? The answer appears to be yes, at least for the first of a female's partners, since the female does not re-mate for days, and the first male to mate with her sires all the eggs she lays in that interim.

Other cantharidiphiles appear to use cantharidin in much the same way. The males alone seem to be the ones that procure the compound, which is destined for transmission to the females and incorporation into the eggs.

Where cantharidiphiles obtain cantharidin is still somewhat of a mystery, although it is generally assumed that the compound is obtained from meloid beetles, the primary source of cantharidin in nature, or from oedemerid beetles, which also produce cantharidin. But the possibility cannot be ruled out that cantharidin is available to cantharidiphiles from entirely different, as yet unknown sources.

REFERENCES

Dettner, K. 1997. Inter- and intraspecific transfer of toxic insect compound cantharidin. *Ecological Studies* 130:115–145.

Eisner, T., S. R. Smedley, D. K. Young, M. Eisner, B. Roach, and J. Meinwald. 1996. Chemical basis of courtship in a beetle (*Neopyrochroa flabellata*): cantharidin as precopulatory "enticing" agent. *Proceedings of the National Academy of Sciences USA* 93:6494–6498.

———— 1996. Chemical basis of courtship in a beetle (*Neopyrochroa flabellata*): cantharidin as "nuptial gift." *Proceedings of the National Academy of Sciences USA* 93:6499–6503.

Görnitz, K. 1937. Cantharidin als Gift und Anlockungsmittel für Insekten. *Arbeiten für physikalische angewandte Entomologie* 4:116–159.

Hilker, M., and T. Meiners. 2002. *Chemoecology of Insect Eggs and Egg Deposition.* Berlin: Blackwell Publishing.

Holz, C. 1995. Die Bedeutung des Naturstoffs Cantharidin bei dem Feuerkäfer *Schizotus pectinicornis* (Pyrochroidae). *Verhandlungen Westdeutscher Entomologentagung* 1994:73–78.

Holz, C., G. Streil, K. Dettner, J. Dütemeyer, and W. Boland. 1994. Intersexual transfer of a toxic terpenoid during copulation and its paternal allocation to developmental stages: quantification of cantharidin in cantharidin-producing oedemerids (Coleoptera: Oedemeridae) and canthariphilous pyrochroids (Coleoptera: Pyrochroidae). *Zeitschrift für Naturforschung* 49c:856–864.

Schütz, C., and K. Dettner. 1992. Cantharidin-secretion by elytral notches of male anthicid species (Coleoptera: Anthicidae). *Zeitschrift für Naturforschung* 47c:290–299.

49

Class INSECTA
Order COLEOPTERA
Family Tenebrionidae
Adelium percatum
A darkling beetle

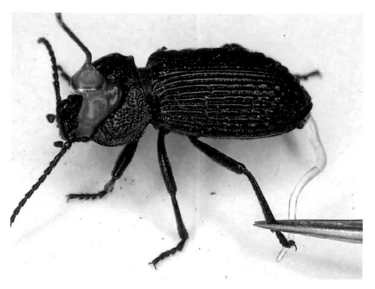

Adelium percatum, everting its left gland in response to pinching of its left hindleg. The beetle is fastened to a wire with a dab of wax.

The Tenebrionidae, or darkling beetles, include some 1,700 genera worldwide, with 18,000 species. Many are chemically protected by defensive glands, present most often in the rear of the abdomen, but sometimes also in the thorax. The glands are of two types: compressible ones that expel their contents as a spray or ooze (see Chapter 51) or eversible ones that expose their contents upon eversion.

In the Australian tenebrionid *Adelium percatum,* the glands are

4. 2,3-Dimethyl-1,4-benzoquinone   16. 2-Methyl-1,4-benzoquinone

118. 1-Pentadecene

abdominal and of the eversible type. They are atypical in that they project to great length when everted. Eversible glands in tenebrionids, as a rule, barely protrude from the abdominal tip when everted (see Chapter 50).

A. percatum can extrude its glands one at a time. Typically, when stimulated on one side, say by the pinching of a leg, it everts the gland from that side only. If pinched briefly, it may evert the gland only in part. But if the pinch is sustained, it eventually everts the gland in its entirety, and by so doing may be able to reach as far as the tip of the midleg or hindleg. The foreleg is just beyond reach. The beetle also is able to exercise some control over the direction of extrusion of its glands. Thus if a hindleg is stimulated at its base, the beetle extrudes the gland at a sharper angle relative to the body than if the leg is stimulated at its tip.

The glandular tubes of *A. percatum,* when extruded, are turgid with contained fluid. We assume this fluid to be blood and that it is through a rise in blood pressure that gland eversion is effected.

The secretion of *A. percatum* contains benzoquinones (4, 16), and the hydrocarbon 1-pentadecene (118). Such mixed production of quinones and hydrocarbons is quite typical for the defensive glands of tenebrionids.

Sharing the natural habitat of *A. percatum* is the closely related *Adelium pustulosum.* This beetle has the same glandular defensive system as *A. percatum,* but differs from the latter in that it also produces a distinctly audible sound when disturbed. The sound is generated by the beetle the moment it is picked up, even before it everts

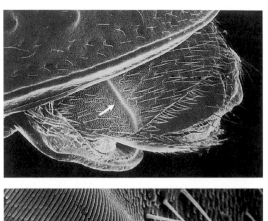

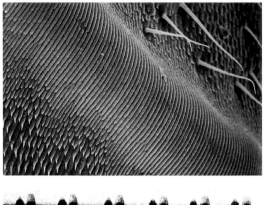

Top: Rear view of *Adelium pustulosum,* showing the left serrated ridge (arrow) projecting from beneath the elytral margin against which it is scraped when the beetle generates its disturbance sound. Middle: An enlarged view of the serrated ridge. Bottom: An acoustical trace of the disturbance sound of *A. pustulosum.* The sound consists of a series of bisyllabic chirps, about 200 milliseconds in duration.

its glands. Both males and females produce the sound, which has the quality of a persistent stridulation of broad frequency range (1 to 60 kilohertz). The sound is engendered by the rhythmic deflection of the abdominal tip. The beetle has two serrated ridges on the last abdominal segment, which it scrapes back and forth across the margins of the wing covers (elytra), thereby generating what is essentially a bisyllabic chirp.

Stridulatory mechanisms are common in insects. Many insects produce sounds for communicative purposes in courtship, but production of sounds in response to disturbances is also well documented. Disturbance calls are emitted, for instance, by some cockroaches and Hemiptera, as well as by long-horned beetles (Cerambycidae) and certain larval and adult Lepidoptera. These insects are sometimes chemically protected, and it has been suggested that their calls are intended to warn predators against inflicting an attack. The sounds could therefore fulfill the same role as gaudy coloration. Many protected insects are brightly colored (see Chapter 62), and there is evidence that predators heed the warning implied by such coloration. The term "aposematism" has been coined to refer to the defensive advertisement effected by visual adornment in protected animals. Defensive sounds could similarly be aposematic (see Chapters 14 and 21). They could be the acoustic counterparts of visual aposematic signals, differing from these only in that they could fulfill their function both in the daylight and in darkness.

It is hard to specify exactly which predators might be on *Adelium*'s enemy list, but both vertebrates and invertebrates could be represented. Benzoquinones are among the most widely occurring components of arthropod defensive secretions (see, for instance, Chapters 2, 8, 15, 16, 17, 21, 35, 50, and 51), and there can be no question that they fulfill a defensive role in *Adelium*. The hydrocarbon in the *Adelium* secretion, 1-pentadecene, could act as a repellent and surfactant.

REFERENCES

Eisner, T., D. Aneshansley, M. Eisner, R. Rutowski, B. Chong, and J. Meinwald. 1974. Chemical defense and sound production in Australian tenebrionid beetles (*Adelium* spp.). *Psyche* 81:189–208.
Masters, W. M. 1979. Insect disturbance stridulation: its defensive role. *Behavorial Ecology and Sociobiology* 5:187–200.
———— 1979. Irradiance modulation used to examine sound-radiating cuticular motion in insects. *Science* 203:57–60.

50

Class INSECTA

Order COLEOPTERA

Family Tenebrionidae

Bolitotherus cornutus

The forked fungus beetle

A shelf fungus of a type commonly inhabited by *Bolitotherus cornutus.*

Tenebrionid beetles are primarily scavengers, feeding on a variety of dead materials, mostly of plant and fungal origin. They occur in many habitats, in deserts, under rocks and logs, in rotting wood, in caves, and in animal burrows. Some species are pests of stored products; others feed on roots of agricultural crops.

Bolitotherus cornutus is an inhabitant of eastern North America,

A *Bolitotherus cornutus* female, in ventral view, with her defensive glands retracted (left) and everted (right).

where it occurs in the shelf fungi that grow on the outside of tree stumps and logs. Adults are active at night, when they feed on the fungal tissue. They hide within the fungus in the daytime. Their enemies are likely to be mostly nocturnal prowlers, including ants, spiders, and rodents. Their common name, forked fungus beetle, derives from the fact that the males have a pair of conspicuous "horns" projecting anteriorly from the pronotum. The horns function in male-male competition and provide a basis for mate selection on the part of the female.

B. cornutus is well protected. The beetles are dark brown and have a roughened, nonglossy body surface, causing them to blend in well with the fungus on which they live. They are encased in a tough exoskleton, and usually "death-feign" when disturbed. When thus feigning, they withdraw their legs into a set of body grooves that serve specifically for the purpose. In common with most tenebrionids, *B. cornutus* has a pair of abdominal defensive glands, which in its case are of the eversible rather than dischargeable kind (see Chap-

16. 2-Methyl-1,4-benzoquinone   **33.** 2-Ethyl-1,4-benzoquinone

The millipede *Orthoporus flavior* coiling in response to being stimulated by puffs of breath.

ter 49). When disturbed, the beetles extrude these glands, turning them inside out and exposing their secretory contents. The secretion has been shown to contain 1,4-benzoquinones (**16, 33**).

An unusual feature of *B. cornutus* is that it may extrude its glands preemptively, in response to the mere anticipation of an attack. All you need to do to cause the beetle to evert its glands is breathe on it. This readiness to deploy its chemical weapons may serve the beetle well, especially in defense against predators such as mice, which could inflict a fatal injury with their very first bite. Experiments have shown that if the gland openings of *B. cornutus* are sealed with glue to prevent the glands from being everted, the beetles are more likely to sustain injury than if they have the option of extruding the glands.

Further experiments showed that the beetle recognizes breath on the basis of specific traits. Simply directing an air stream onto the beetle does not elicit gland eversion. For eversion to occur, the stream needs to be humid, warm, and pulsed—truly breathlike. It would make little sense for the beetle to respond indiscriminately to air movement of any sort. There is a cost associated with the production of the defensive quinones, and thus there is a good reason to restrict the use of the glands to real emergencies.

Most arthropods are programmed to refrain from activating their chemical weapons unless directly contacted by an enemy. By responding preemptively, *B. cornutus* is therefore unusual, though not unique. The desert millipede *Orthoporus flavior,* for instance, coils instantly in response to puffs of breath, and if breathed upon persistently proceeds to discharge its glands. The millipede *Glomeris marginata* (see Chapter 11) likewise responds by coiling when stimulated by puffs of breath.

REFERENCES

Conner, J. 1989. Density-dependent sexual selection in the fungus beetle, *Bolitotherus cornutus. Evolution* 43:1378–1386.

——— 1989. Older males have higher insemination success in a beetle. *Animal Behavior* 38:503–509.

Conner, J., S. Camazine, D. Aneshansley, and T. Eisner. 1985. Mammalian breath: trigger of defensive chemical response in a tenebrionid beetle *(Bolitotherus cornutus). Behavioral Ecology and Sociobiology* 16:115–118.

51

Class INSECTA
Order COLEOPTERA
Family Tenebrionidae
Eleodes longicollis
A darkling beetle

Eleodes longicollis in its defensive posture.

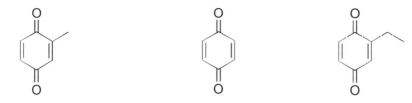

16. 2-Methyl-1,4-benzoquinone   **30.** 1,4-Benzoquinone   **33.** 2-Ethyl-1,4-benzoquinone

119. 1-Nonene   **120.** 1-Undecene

121. 1-Tridecene   **2.** Caprylic acid

The color black, you would imagine, is impractical in a desert setting, since it offers little protection against the heat. Yet as naturalists are quick to note, black insects, mainly beetles, abound in desert
settings. In the deserts of the southwestern United States there exist
a multitude of jet-black beetles, ranging from mid-size to large and
belonging primarily to the family Tenebrionidae. These beetles are
quiescent in midday when the sun is most intense and are active primarily at dusk and dawn, in the twilight, when they are on the
march to or from their feeding sites. True color is poorly perceived
in dim light, and there is no better way to be noticeable on sand in a
crepuscular setting than by being black. Add to this the fact that
dawn and dusk are the times when vertebrates are on the prowl, and
you are faced with the apparent paradox that black desert beetles are
most conspicuous when they are most at risk.

The truth is, however, that desert tenebrionids are for the most
part chemically protected and that it is to their advantage to advertise themselves. Being black in a desert setting may be the best way
for a beetle to tell a predator: "Watch out, I'm a mouthful you're

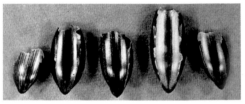

Left: *Eleodes longicollis,* dissected open to reveal the two large defensive glands in the rear of its abdomen. Top right: A grasshopper mouse *(Onychomys torridus)* feasting on *E. longicollis.* Bottom right: Remnants of *E. longicollis* found in grasshopper mouse territory in the desert.

likely to regret." Notable among North American desert tenebrionids is *Eleodes longicollis,* a large, shiny, spectacularly beautiful black beetle. *E. longicollis,* like all members of its genus, has a pair of large abdominal glands from which it discharges a foul-smelling, irritating secretion when disturbed. The fluid contains 1,4-benzoquinones (**16, 30, 33**), plus several hydrocarbons (**119–121**) and caprylic acid (**2**). The quinones are doubtless the major repellents in the mixture, while the other components may serve as solvents of the quinones and as spreading and penetration-promoting agents. The beetle has a distinctive behavior when disturbed, which it shares with many other members of its genus. No sooner is it touched than it comes to a halt and assumes a headstand. It presses its head against the ground and, by straightening out its hindlegs, points its rear skyward. It typically remains motionless in that stance, ready to

Two glandless, headstanding mimics of *Eleodes,* an unidentified species of the tenebrionid genus *Gonasida* (left) and the cerambycid *Moneilema appressum* (right).

use its defensive glands should the disturbance persist. Glandular discharges may be forcible, and involve actual spraying of secretion, but they may also involve no more than the emission of a droplet that clings to the abdominal tip and that the beetle attempts to spread by use of its hindlegs.

The secretion is highly effective against ants, which in a desert setting are a major hazard, and it is doubtless deterrent also to other predators, including rodents and birds. It is, however, ineffective against one species of rodent, the grasshopper mouse, *Onychomys torridus,* which has the remarkable habit of holding the beetle upright when it comes upon it and forcing its rear into the sand, thereby causing the secretion to be discharged ineffectually into the soil. The mouse commences eating the moment it has a beetle in its hold and eventually consumes almost the entire prey, leaving only the abdominal tip (with the intact glands) and the legs and wing covers (which offer little nutrients). Such remnants are a frequent sight in desert areas where grasshopper mice abound.

"Headstanding" is not restricted to *Eleodes,* but occurs also in other beetles. One striking headstander is *Moneilema appressum,* a member of the longhorn beetle family (Cerambycidae). This insect is jet black and wingless like *E. longicollis* and, like the latter, is active on the desert floor. It too "freezes" and raises its rear when touched, although it has no defensive glands. *Moneilema* is evidently a mimic of *Eleodes,* a mimic of the Batesian kind, intrinsically defenseless and reliant on being confused with the "real thing." Some tenebrionids (for example, the species of the genus *Gonasida*) pull essentially the same trick. They lack defensive glands, but stand on their heads like *E. longicollis* when threatened, thereby presumably holding predators at bay.

REFERENCES

Chadha, M. S., T. Eisner, and J. Meinwald. 1961. Defense mechanisms of arthropods. IV. *Para*-benzoquinones in the secretion of *Eleodes longicollis.* Lec. (Coleoptera: Tenebrionidae). *Journal of Insect Physiology* 7:46–50.

Eisner, T. 1966. Beetle's spray discourages predators. *Natural History* 75:42–47.

Eisner, T., F. McHenry, and M. M. Salpeter. 1964. Defense mechanisms of arthropods. XV. Morphology of the quinone-producing glands of a tenebrionid beetle (*Eleodes longicollis* Lec.). *Journal of Morphology* 115:355–400.

Hurst, J. J., J. Meinwald, and T. Eisner. 1964. Defense mechanisms of arthropods. XII. Glucose and hydrocarbons in the quinone-containing secretion of *Eleodes longicollis. Annals of the Entomological Society of America* 57:44–46.

Meinwald, Y. C., and T. Eisner. 1964. Defense mechanisms of arthropods. XIV. Caprylic acid: an accessory component of the secretion of *Eleodes longicollis. Annals of the Entomological Society of America* 57:513–514.

Meinwald, J., K. F. Koch, J. E. Rogers, Jr., and T. Eisner. 1966. Bio-synthesis of arthropod secretions. III. Synthesis of simple *p*-benzoquinones in a beetle *(Eleodes longicollis). Journal of the American Chemical Society* 88:1590–1592.

52

Class INSECTA
Order COLEOPTERA
Family Scarabaeidae
Trichiotinus rufobrunneus
A scarab beetle

Trichiotinus rufobrunneus.

The large family Scarabaeidae (2,000 genera, 25,000 species) includes a broad diversity of beetles. Many are dung or carrion feeders. Others feed on plant matter or fungi. Some inhabit ant or termite colonies, while others frequent the burrows of vertebrates.

Given their great evolutionary success, it is surprising that so few scarabs are chemically defended. They usually have a tough body armor, but as a rule lack defensive glands. If you took an assortment of scarabs, such as are commonly attracted to a light trap on a summer

A rear-end view of *Trichiotinus rufobrunneus,* displaying its wasp-like appearance.

night, and offered them to a thrush, a blue jay, or a mouse, chances are you would find most of them acceptable to these predators.

› There are, of course, exceptions. Among scarabs known as dung beetles, for instance, there are some that emit defensive fluids from the rear of the abdomen when disturbed. In *Canthon augustatus,* a beetle associated with monkey dung in Panama, the abdominal effluent is glandular in origin and distinctly different in odor in males and females. Whether in this beetle the secretion has a dual defensive and pheromonal role remains open to question. The composition of the fluid is unknown.

Trichiotinus rufobrunneus is an unusual scarab beetle in that it seems to derive protection from its appearance. The beetle occurs in central Florida, where it is found on various flowers, principally those of *Opuntia* cacti. It typically sits facing inward on these flowers, so that it is visible from the rear. It then has an uncanny resemblance to a wasp. The tip of its abdomen has the triangular shape of a wasp head and bears a pair of "eyes" resembling those of a wasp.

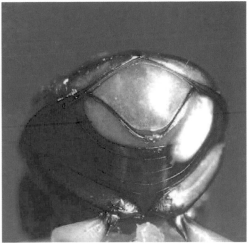

Left: A pair of dung beetles *(Canthon augustatus)* from Panama, rolling a dung ball. Right: An end-on view of the abdominal tip of *C. augustatus,* showing whitish secretion oozing from the right defensive gland.

The "retro-stare" might help *T. rufobrunneus* in two contexts. It could discourage competing pollinators from landing on the flower and wasp-shunning predators from making an approach. Other species of *Trichiotinus,* as for instance *T. lunulata,* that coexist with *T. rufobrunneus* in Florida, have a similar wasp-like rear.

REFERENCES

Cott, H. B. 1957. *Adaptive Coloration of Animals,* 2nd ed. London: Methuen.
Wickler, W. 1968. *Mimicry in Plants and Animals.* London: Weidenfeld and Nicolson.

53

Class INSECTA
Order COLEOPTERA
Family Chrysomelidae
Hemisphaerota cyanea
A tortoise beetle

Hemisphaerota cyanea on a palmetto frond.

Ants, even when singly on the prowl, can be a considerable threat to small insects. Many have mandibles strong enough to seize an individual insect and carry it off to the nest. Insects may have defensive glands to fend off such attacks, or the ability to leap or fly away. The chrysomelid beetle *Hemisphaerota cyanea* is unusual in that it stays put when attacked. It has a remarkable ability to cling with its feet and it puts this ability to use whenever it is attacked by ants.

Hemisphaerota cyanea withstanding a pull of 2 grams.

H. cyanea is a member of the chrysomelid subfamily Cassidinae, which includes the tortoise beetles. The name is appropriate. Tortoise beetles bear a shield, made up of the pronotum and elytra (wing covers), which covers their entire dorsal surface, giving them the appearance of miniature turtles. The shield provides cassidines with a first line of defense because it renders the beetles hard to grasp. This is particularly true for *H. cyanea,* which is of nearly perfect hemispherical shape and has nothing protruding from its surface that might be grasped by an ant. Its legs and antennae are ordinarily concealed beneath the shield.

H. cyanea clings with its feet by using an oil. The feet, or tarsi, are miniature brushes, each bearing a dense covering of bristles on its sole. The bristles are forked at the tip, each ending in two pads that are wetted with oil and brought in contact with the substrate when the foot is pressed down. Because they are forked, the bristles hold their parallel alignment during foot touchdown, without being deflected to one side or the other. The oil provides the means by which

Left: A formicine ant attempting unsuccessfully to seize a *Hemisphaerota cyanea.*
Right: A ventral view of a *H. cyanea* showing the large yellow feet.

the pads stick to the substrate. When the pads are pressed against the substrate, the oil on their tips is squeezed into a thin film, which is all that is needed to fasten the pads in place. The adherence is similar functionally to that provided by a droplet of water squeezed into a thin film between two plates of glass. The oil is secreted from tiny glands that open between the bases of the bristles. It creeps by capillarity to the tips of the bristles, wetting these each time the foot is lifted up and readying them for the next touchdown.

The strength of the bond between foot and substrate is considerable, largely because of the numbers of bristles involved. There are some 10,000 bristles per foot in *H. cyanea,* or 120,000 contact pads per beetle. You can measure the strength of the adhesion that *H. cyanea* is able to secure when pressed down by attaching the beetle with a hook to an electronic force transducer and applying a pull. The value you would obtain is an impressive 0.8 grams, or nearly 60 times the beetle's weight, which the animal can withstand for up to 2 minutes. This appears to be longer than the time individual ants are willing to invest in pulling on the beetle. Pulls in excess of 0.8 grams are withstood by the beetle for shorter periods. The maximum force the beetle can withstand, albeit for only a few seconds, is on the order of 3 grams.

To detach itself, the beetle rolls its feet off the substrate in the

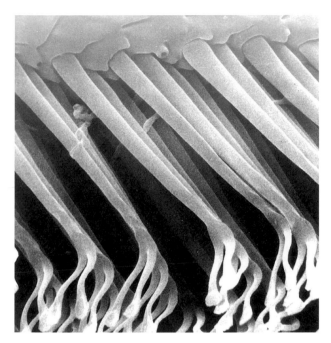

An enlarged view of the foot bristles of *Hemisphaerota cyanea*. Note that the bristles are forked at the tip. A single oil pore is visible at the base of a bristle.

same manner that you would peel a strip of adhesive tape from a surface. The feet consist of three articulated segments that lend themselves to being detached in sequence.

Defense is costly to *H. cyanea* because the beetle loses oil every time it is attacked. Wherever one of its tiny foot pads touches down during an attack, a quantity of oil is deposited. The total amount of oil relinquished when all six legs are committed to contact is substantial, but the beetle is evidently able to bear the cost. Interestingly, during ordinary locomotion, the beetle does not set its feet down flat, but brings into contact only a fraction of the bristles of each foot. By thus "tiptoeing" it cuts down on the quantity of oil lost.

Possession of bristly feet is commonplace among the Chrysomelidae, but most have far fewer bristles per foot than *H. cyanea* and can therefore withstand only relatively moderate pulling forces. De-

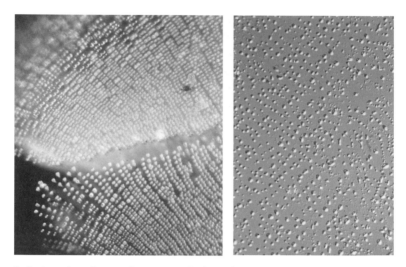

Left: An enlarged view of a portion of a foot of *Hemisphaerota cyanea*. Right: Oil droplets left on the substrate where a foot touched down.

fense in their case is achieved by means other than foot adhesion. Some tortoise beetles rely primarily on tarsal claws to attach themselves when attacked. *H. cyanea* lives on leaves too tough to be punctured by claws. The other species reside on succulent leaves that are readily pierced by tarsal hooks.

Although well protected by its feet, *H. cyanea* does have an enemy, a reduviid bug, commonly known as the wheel bug *(Arilus cristatus)*. This predator injects a paralyzing fluid into *H. cyanea,* thereby preventing the beetle from tightening the muscles that it uses to press down its feet.

The tarsal oil of *H. cyanea* has been characterized and found to consist of a mixture of long-chain saturated and unsaturated hydrocarbons.

REFERENCES

Artzt, E., S. Gorb, and R. Spolenak. 2003. From micro to nano contacts in biological attachment devices. *Proceedings of the National Academy of Sciences USA* 100:10603–10606.

Attygalle, A. B., D. J. Aneshansley, J. Meinwald, and T. Eisner. 2001. Defense by foot adhesion in a chrysomelid beetle (*Hemisphaerota cyanea):* characterization of the adhesive oil. *Zoology* 103:1–6.

Eisncr, T., and D. J. Aneshansley. 2000. Defense by foot adhesion in a beetle (*Hemisphaerota cyanea*). *Proceedings of the National Academy of Sciences USA* 97:6568–6573.

Jolivet, P. H., and M. L. Cox, eds. 1996. *Chrysomelidae Biology,* vol.2: *Ecological Studies.* Amsterdam: SPB Academic Publishing.

Jolivet, P. H., M. L. Cox, and E. Petitpierre, eds. 1994. *Novel Aspects of the Biology of Chrysomelidae.* Dordrecht, Netherlands: Kluwer.

Pennisi, E. 2002. Biology reveals new ways to hold on tight. *Science* 296:250–251.

54

Class INSECTA

Order COLEOPTERA

Family Chrysomelidae

Gratiana pallidula

A tortoise beetle

Gratiana pallidula on *Solanum eleagnifolium.*

Feces, to most animals, are in the category of hazardous wastes. Feces may carry parasites or pathogens and are for that reason best deposited where one is not likely to encounter them again. But there are insects that put feces to use. An example is provided by the tortoise beetles, members of the chrysomelid subfamily Cassidinae, which as larvae use their own feces to construct a defensive packet that they carry over their back. To accommodate the packet the lar-

Top left: A young larva of *Gratiana pallidula*. Top right: The same larva when older. Bottom left: The same larva when it was about to pupate; the fecal package has been picked off with forceps to expose the two-pronged fecal fork. Bottom right: A larva of *Chelimorpha cassidea* with its pasty shield.

vae have a two-pronged fork, extending forward from just above the anus to nearly as far as the head. One such packet-carrying larva is that of *Gratiana pallidula,* a tortoise beetle from the southwestern United States, commonly found on *Solanum eleagnifolium,* a plant of the nightshade family (Solanaceae).

The *G. pallidula* larva begins feeding as soon as it hatches, and it deposits feces on the fork from the very start. When it molts, it does not discard the cast cuticle, but adds it to the feces on the fork. The feces are not lost as the larva grows, but are retained from one larval stage to the next, so that the packet eventually comes to be big enough to cover most of the larva.

The larva uses the packet as a maneuverable shield. By twisting and flexing its rear, it can tilt the shield in any direction, and it relies

Top: The soft-bodied larva of *Hemisphaerota cyanea* is ordinarily totally hidden from view (left); lifting off the thatch exposes the larva (right). Bottom: The predaceous carabid beetle *Calleida viridipennis* forcing its way into an *H. cyanea* thatch.

on this capability to block assaults. If, for example, the larva is poked with a probe, it rotates the shield in such a manner as to interpose it between itself and the probe. If it is touched on a flank, it tilts the shield to the side. If it is stimulated in the front or in the rear, it raises the rear and tilts the shield forward, or presses the rear down and tilts the shield backward. The action is effective against small predators such as ants and spiders.

A similar shield is constructed by the larva of *Chelimorpha cassidea*. This animal, which feeds on succulent plants, produces pasty feces, with the result that its shield is itself pasty. Whether the shield is rendered chemically noxious by virtue of its wetness, and therefore defensively more effective, has not been established for *C. cassidea*. But in certain other cassidine species, the shield has indeed been shown to contain chemical repellents derived from the food-plant.

An unusual shield or thatch is produced by the larva of *Hemisphaerota cyanea* (see Chapter 53). This larva feeds on palmetto plants (for example, *Serenoa repens* and *Sabal palmetto*) and produces feces in the form of long strands. It has the caudal fork characteristic of cassidines, although in its case the fork's prongs are relatively short. The larva attaches the fecal strands to the fork, one after the other, as they are squeezed from the anus. It imparts to each strand a definite curvature, so that half are bent to fit on the left of the thatch and the other half to fit on the right. It produces the two types of strands in regular alternation, thereby ensuring that the thatch maintains overall symmetry as it grows. It attaches the strands to the fork with a special glue that it secretes from glands positioned just above the anus. The strands are dry and the thatch that the larva constructs with them forms what, to many predators, is an impenetrable barrier. Insectivorous coccinellid larvae ignore *H. cyanea* larvae when they come upon the thatch, and so do ants. Yet, one carabid beetle, *Calleida viridipennis,* is in no way discouraged by the thatch. This beetle forces its way, or chews its way into the thatch, and eats the larva when it comes upon it. It rejects only the thatch and the larva's anal turret.

REFERENCES

Eisner, T., and M. Eisner. 2000. Defensive use of a fecal thatch by a beetle larva (*Hemisphaerota cyanea*). *Proceedings of the National Academy of Sciences USA* 97:2632–2636.

Eisner, T., E. van Tassell, and J. E. Carrel. 1967. Defensive use of a "fecal shield" by a beetle larva. *Science* 158:1471–1473.

Morton, T. C., and F. V. Vencl. 1998. Larval beetles form a defense from recycled host-plant chemicals discharged as fecal wastes. *Journal of Chemical Ecology* 24:765–785.

Mueller, C., and M. Hilker. 1999. Unexpected reactions of a generalist predator towards defensive devices of cassidine larvae (Coleoptera, Chrysomelidae). *Oecologia* 118:166–172.

Olmstead, K. L., and R. F. Denno. 1993. Effectiveness of tortoise beetle larval shields against different predator species. *Ecology* 74:1394–1405.

Vencl, F. V., T. C. Morton, R. O. Mumma, and J. C. Schultz. 1999. Shield defense of a larval tortoise beetle. *Journal of Chemical Ecology* 25:549–566.

55

Class INSECTA

Order COLEOPTERA

Family Chrysomelidae

Plagiodera versicolora

The imported willow leaf beetle

Plagiodera versicolora on a willow leaf.

Chemical defenses occur in all developmental stages of insects, including the larval stage. Among caterpillars alone, for instance, protection may be provided by dischargeable or eversible glands (see Chapters 58 and 64), urticating hairs (see Chapter 63), or systemic toxins (see Chapter 62).

Among beetle larvae defenses are also diverse, and may take the form of fecal shields (see Chapter 54), systemic toxins (see Chapter

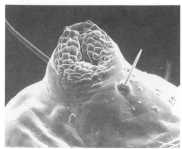

Left: A *Plagiodera versicolora* larva responding to pinching with forceps by emitting droplets of defensive secretion from its eversible glands. Right: A single gland of *P. versicolora* in its uneverted (top) and everted condition (bottom); the secretion was washed away with solvent in the course of readying the preparation for scanning electronmicroscopy.

46), or eversible glands. One group of beetles possessing eversible glands in the larval stage are those of the tribe Chrysomelini, of the chrysomelid subfamily Chrysomelinae. *Plagiodera versicolora* is a case in point. This small, dark blue beetle occurs commonly on willow trees in the eastern United States. Its larva has nine pairs of glands, arranged segmentally along the sides of its body. The glands consist of small pouches replete with fluid, ordinarily kept withdrawn, but promptly everted when the larva is prodded or pinched.

The *P. versicolora* larva grades its response in accord with the intensity of the stimulus. If the larva is seized bodily with forceps and lifted off the substrate, it is likely to evert all glands at once. If, however, it is subjected to a localized stimulus, such as the pinching of a leg, it may restrict its response to the eversion of the glands closest to that leg. The eversion itself may be of the briefest duration and involve no more than the momentary "airing" of the exposed secretion. Typically, following cessation of the stimulus, the glandular

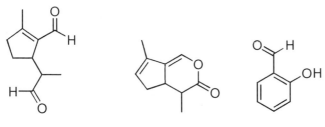

122. Chrysomelidial   123. Plagiolactone   124. Salicylaldehyde

sacs, together with the extruded droplets, are pulled back into the body. Loss of secretion is thereby held to a minimum.

The secretion of *P. versicolora* is potently repellent to arthropod predators. It was found to contain two previously unknown compounds that have been named chrysomelidial (**122**) and plagiolactone (**123**). Both compounds are methylcyclopentanoid monoterpenes, a group of isoprenoids of known anti-insectan potency, which include compounds previously isolated from insects and plants (see Chapters 20, 38, and 39). Interestingly, the *P. versicolora* secretion, in addition to deterring predators, may have a repellent effect on competing herbivores that feed on the same plant.

Related chrysomelids have larval glands structurally similar to those of *P. versicolora*, but of different secretory capacity. In *Chrysomela scripta*, for instance, the larvae secrete a fluid rich in salicylaldehyde (**124**), a compound also strongly repellent to insects. In *C. scripta*, the larval glands actually protect the pupa as well. When the larvae pupate, they shed their entire exoskeletal cuticle, and with it the linings (and therefore secretory contents) of their glands. They do not cast away this cuticle, however, but retain it as a loosely fitting cover over the rear of the pupal abdomen. The larval glands with their contents thus come to lie in the space between the cast cuticle and the pupal abdominal wall. If the pupa is disturbed by being prodded in some fashion, it rotates the abdomen, thereby squeezing the glandular pouches and forcing some of the secretion to the outside. Predators that attack the pupae are thus likely to be exposed to the larval secretion. Other members of the tribe Chrysomelini appear to make similar use of the larval secretion in the pupal stage.

Chrysomela scripta on a willow leaf.

Left and middle: A *Chrysomela scripta* larva responding to pinching of a leg by everting, first, the glands from the side of the leg stimulated and then, in addition, the glands on the opposite side. Right: A *C. scripta* larva being eaten by a carnivorous pentatomid bug. The predator can impale the larva on its beak with minimal risk of being contacted by the prey's secretion.

Left: A pupa of *Chrysomela scripta* projecting from the larval cuticle that it has only partially shed. The white larval secretion is visible in the glandular pouches that are a part of the larval cuticle. Right: A *Formica* ant being wetted by the larval secretion in the course of an attack upon a *C. scripta* pupa.

The glandular substances produced by larvae of the Chrysomelini may in some cases be synthesized by the larvae themselves from closely related chemical precursors present in the diet. Thus salicylaldehyde-producing larvae that feed on willow or poplar trees (family Salicaceae) may synthesize the aldehyde from salicin, a phenylglucoside found in the leaves of the host plant.

REFERENCES

Hinton, H. E. 1951. On a little known protective device of some chrysomelid pupae (Coleoptera). *Proceedings of the Royal Entomological Society London* 98:449–472.

Kearsley, M. J. C., and T. G. Whitham. 1992. Guns and butter: a no cost defense against predation for *Chrysomela confluens. Oecologia* 92:556–562.

Meinwald, J., T. H. Jones, T. Eisner, and K. Hicks. 1977. New methylcyclopentanoid terpenes from the larval defensive secretion of a chrysomelid beetle *(Plagiodera versicolora). Proceedings of the National Academy of Sciences USA* 74:2189–2193.

Pasteels, J. M., R. M. Rowell-Rahier, J. C. Braekman, and A. Dupont. 1983. Salicin from host plant as precursor of salicyl aldehyde in defensive secretion of chrysomeline larvae. *Physiological Entomology* 8:307–314.

Rank, N. E., and J. T. Smiley. 1994. Host-plant effects on *Parasyrphus melanderi* (Diptera: Syrphidae) feeding on a willow leaf beetle *Chrysomela aeneicollis* (Coleoptera: Chrysomelidae). *Ecological Entomology* 19:31–38.

Raupp, M. J., F. R. Milan, P. Barbosa, and B. A. Leonhardt. 1986. Methylcyclopentanoid monoterpenes mediate interactions among insect herbivores. *Science* 232:1408–1410.

56

Class INSECTA
Order LEPIDOPTERA
Family Dalceridae
Dalcerides ingenita
A dalcerid moth

A *Dalcerides ingenita* larva.

Caterpillars are eating machines. Their soft integument allows them to accommodate the increasing bulk of their organs as they build up mass between molts, and it also provides them with the wormlike flexibility that is basic to their crawling life style. Being soft-bodied also means that they are relatively easily pierced. If torn open by a predator they lose body fluids and go limp, usually with fatal consequences. Not surprisingly, caterpillars have evolved a multiplicity of defenses. Some avoid detection by being camouflaged or by being active strictly at night, while a great many derive protection from venomous spines, defensive glands, vomited fluid, slimy body coatings, or systemic toxins. Many defended caterpillars are warningly colored. Undefended ones may be aposematic as well, and benefit from being mistakenly identified as being noxious (see Chapters 57–65 for examples of caterpillar defenses).

Slimy integumental coverings are rare in arthropods, and appear to be restricted to species that have a thin exoskeleton. In the sawfly larva *Caliroa cerasi* (Hymenoptera, family Tenthredinidae), the sluglike body is covered in its entirety with a viscous liquid that is presumed to act as a deterrent to such predators as ants.

Among the Lepidoptera, a sticky integumental coating is present in larvae of the Dalceridae, a family of some 85 species, mostly from the New World tropics. Only one species, *Dalcerides ingenita,* is native to the United States. The caterpillar of this moth is sluglike, and bears a thick dorsal investiture that looks warty and feels sticky. The coating can be easily removed from the larva. Tugging at a single wart with forceps typically results in detachment of a whole string of warts.

Encounters between *D. ingenita* larvae and ants, staged in the laboratory, showed that the slimy coating provides effective defense. Ants made repeated contact with the larvae, but in most cases withdrew without attempting to bite. In the few cases where bites occurred, the ants were noticeably affected. They either became stuck to the investiture and had to struggle to free themselves, or they pulled away, visibly encumbered with dabs of coating. Contaminated ants were slow to clean themselves and tended therefore not to attack again. They did, however, eventually recover.

Chemical analysis showed that the integumental coating of *D. ingenita* is devoid of components that could be repellent or irritant to predators. The coating, therefore, appears to act strictly as a mechanical encumbering agent.

Among the Lepidoptera, only certain species of the Limacodidae, a family closely related to the Dalceridae, are known also to have sticky integumental coatings.

A *Dalcerides ingenita* larva. Top left: Close-up view of the sticky cover, showing the "warts." Top right: A portion of the sticky cover being pulled off with forceps. Bottom: A formicine ant biting into a larva and then withdrawing after its mouthparts became gummed up.

REFERENCES

Brower, L. P. 1989. Chemical defense in butterflies. In R. I. Vane-Wright and P. R. Ackery, eds., *The Biology of Butterflies.* Princeton: Princeton University Press.

Epstein, M., S. R. Smedley, and T. Eisner. 1994. Sticky integumental coating of a caterpillar (Dalceridae): a deterrent to ants? *Journal of the Lepidopteran Society* 48:381–386.

Stehr, F. W., and N. McFarland. 1987. Dalceridae (Zygaenoidea). In F. W. Stehr, ed., *Immature Insects.* Dubuque: Kendall/Hunt.

57

Class **INSECTA**

Order **LEPIDOPTERA**

Family Noctuidae

Litoprosopus futilis

The palmetto borer moth

Eggs of *Litoprosopus futilis.* The protective cover of scales is provided by the mother moth.

Whas a lion makes a kill, it does not eat the entire carcass. It deliberately avoids ingesting the contents of the intestine. Squeezing the organ as if it were a tube of toothpaste, it empties the intestine of its fill, and only then proceeds to eat the fleshy wall. Intestinal contents, as a rule, are excluded from a predator's meal, and are in fact so generally repellent that some animals have evolved means for putting their own feces to defensive use.

Enteric fluids can be voided both fore and aft, and the two options are not mutually exclusive. Insects in particular are prone to use gut contents for defense. Some vomit when attacked, others defecate, while still others do both at the same time.

Insects that vomit upon disturbance include beetles, fly larvae, and larval hymenopterans, but the habit is most widespread among grasshoppers (see Chapter 21) and caterpillars. It is virtually impossible to pick up a grasshopper without causing it to void its crop contents, and the same holds for many caterpillars. Both grasshoppers and caterpillars are herbivores, and their oral effluent can therefore be expected to be "spiked" to varying degrees with deterrent chemicals stemming from their foodplants. Cases of caterpillars putting noxious plant substances to such secondary use have been documented. In certain Australian caterpillars of the genus *Myrascia* (family Oecophoridae) that feed on eucalypts, the oral effluent consists very largely of repellent oil from the host plant (see Chapter 67). Similarly, tent caterpillars (*Malacosoma americanum,* family Lasiocampidae), when feeding on cyanogenic plants, may derive advantage from the presence in their regurgitant of some of the plant's cyanogenic factors.

Evidence for the effectiveness of the regurgitant has been obtained with the caterpillar of the noctuid moth, *Litoprosopus futilis,* a native of the southeastern United States that feeds on saw palmetto *(Serenoa repens).* The larvae hide in the daytime and feed at night on the unopened floral buds. They vomit readily when provoked. If the effluent is collected and presented at close range in the form of a suspended droplet to a group of ants feeding at a sugar source, the ants will show a distinct avoidance reaction. Wolf spiders tend also to react aversively to the fluid. The regurgitant, besides being repellent, is also irritating. When applied topically to cockroaches, it induces scratching behavior.

L. futilis is able to regurgitate even while in the process of emerging from the egg. If poked with a probe at such a time, it responds by biting and delivering regurgitant. The larva is evidently armed

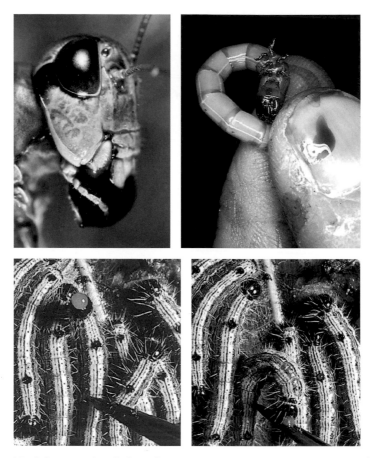

Top left: An unidentified grasshopper regurgitating. Top right: An unidentified tenebrionid beetle larva regurgitating. Bottom: A notodontid caterpillar *(Clostera inclusa)* responding to pinching by revolving its front end and delivering oral effluent upon the forceps used to inflict the pinch.

and ready to put its oral fluid to use before it has even begun feeding on its foodplant.

The use of enteric products for defense can also backfire for a caterpillar. It has been shown that certain salivary components of caterpillars can act to attract wasps that parasitize these caterpillars. One such salivary component is a substance called volicitin (**125**), which appears to have evolved initially as a defensive agent repellent to predators, but may, secondarily, to the detriment of the

A *Litoprosopus futilis* larva being poked as it is emerging from the egg. The larva is itself biting (left) and emitting enteric fluid (right; note the wetted tip of the forceps).

125. Volicitin

producer, have come to serve as a cue that enemy wasps use to find their hosts.

L. futilis belongs to the family Noctuidae, which with some 20,000 species worldwide, is one of the largest families of moths. Within the Noctuidae, *L. futilis* belongs to the subfamily Catocalinae.

REFERENCES

Bowers, M. D. 1993. Aposematic caterpillars: life-styles of the warningly colored and unpalatable. In N. E. Stamp and T. M. Casey, eds., *Caterpillars: Ecological and Evolutionary Constraints on Foraging.* New York: Chapman and Hall.
Brower, L. P. 1984. Chemical defense in butterflies. In R. I. Vane-Wright and P. R. Ackery, eds., *The Biology of Butterflies,* Symposium of the

Royal Entomological Society of London, number 11. London: Academic Press.

Common, I. F. B., and T. E. Bellas. 1977. Regurgitation of host-plant oil from a foregut diverticulum in the larvae of *Myrascia megalocentra* and *M. bractaetella* (Lepidoptera: Oecophoridae). *Journal of the Australian Entomological Society* 16:144–147.

Peterson, S. C., N. D. Johnson, and J. L. LeGuyader. 1987. Defensive regurgitation of allelochemicals derived from host cyanogenesis by eastern tent caterpillars. *Ecology* 68:1268–1272.

Smedley, S. R., E. Ehrhardt, and T. Eisner. 1993. Defensive regurgitation by a noctuid moth larva *(Litoprosopus futilis). Psyche* 100:209–221.

Tumlinson, J. H., W. J. Lewis, and L. E. M. Vet. 1993. How parasitic wasps find their hosts. *Scientific American* March: 100–106.

58

Class INSECTA

Order LEPIDOPTERA

Family Notodontidae

Schizura unicornis

The unicorn caterpillar moth

A *Schizura unicornis* larva.

The ability to eject defensive secretions as sprays evolved independently in many arthropods, including whipscorpions, cockroaches, earwigs, termites, walkingsticks, hemipterans, beetles and ants (see Chapters 1, 13, 15–20, 22, 34, 35, 38, 51, and 68). It also evolved in certain caterpillars, notably in some species of the family Notodontidae. These caterpillars spray from the front end, from a saclike gland situated ventrally just behind the head and opening by way of

68. Formic acid   **1.** Acetic acid   **126.** 2-Tridecanone

127. 2-Pentadecanone   **128.** 2-Tridecyl formate

129. 2-Pentadecyl formate

a slit in the neck region. One such caterpillar is that of *Schizura unicornis.*

The larva of *S. unicornis* is irregularly colored in green and brown, and bizarre in shape. Its dorsal toothlike projections give it a jagged profile that enables it to escape detection, blending in with the margins of partially eaten leaves. The caterpillar occurs in the northeastern United States, and feeds on a diversity of broadleaf trees and shrubs.

S. unicornis aims its spray. When disturbed, as when grasped with forceps, the larva revolves the front end, points the head to where it is seized, and discharges. Predators are thus likely to be hit full blast.

The primary component of the *S. unicornis* secretion is formic acid (**68**). Present in lesser amounts are acetic acid (**1**) and a number of straight-chain lipophilic compounds, including 2-tridecanone (**126**), 2-pentadecanone (**127**), 2-tridecyl formate (**128**), and 2-pentadecyl formate (**129**). The lipophilic components could serve to

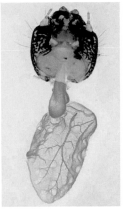

Left: A *Schizura unicornis* larva placed on indicator paper, discharging an aimed jet of spray in response to being pinched in the rear with forceps. Right: The head of *Schizura leptinoides* larva with the defensive gland attached. The gland of *S. leptinoides* is identical in structure to that of *S. unicornis*.

promote the spread and penetration of the fluid on the target, and could thereby quicken the action of the secretion and enhance its irritancy.

Dischargeable glands that emit sprays in arthropods are most often located either at the front or rear of the animal. This makes sense, given that these are the regions of the body that can be flexed or rotated to some extent. Indeed, insects that possess glands that open in the mid-region of the body, such as, for example, Hemiptera (see Chapter 22) or cockroaches (see Chapters 13, 15, 16), are not nearly so effective in aiming their spray as, for instance, carabid beetles (see Chapters 34, 35) or silphid beetles (see Chapter 38), which "fire" from the abdominal tip. Earwigs may be an exception (see Chapter 17). Their glands open near the base of the abdomen, but in their case the abdomen is so flexibly attached to the thorax that they can point their gland openings in most directions.

The family to which *S. unicornis* belongs, the Notodontidae, contains about 650 genera, and more than 2,000 described species.

REFERENCES

Attygalle, A. B., S. R. Smedley, J. Meinwald, and T. Eisner. 1993. Defensive secretion of two notodontid caterpillars *(Schizura unicornis, S. badia). Journal of Chemical Ecology* 19:2089–2104.

Detwiler, J. D. 1922. The ventral prothoracic gland of the red-humped apple caterpillar (*Schizura concinna* Smith and Abbott). *The Canadian Entomologist* 54:176–191.

Poulton, E. B. 1887. The secretion of pure formic acid by lepidopterous larvae for the purpose of defence. *The British Association for the Advancement of Science Report* 5:765–766.

Weatherston, J., J. E. Percy, L. M. MacDonald, and J. A. MacDonald. 1979. Morphology of the prothoracic defensive gland of *Schizura concinna* (J. E. Smith) (Lepidoptera: Notodontidae) and the nature of its secretion. *Journal of Chemical Ecology* 5:165–177.

59

Class INSECTA

Order LEPIDOPTERA

Family Thyrididae

Calindoea trifascialis

A thyridid moth

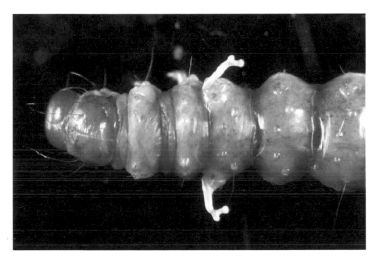

The front end of a larva of *Calindoea trifascialis,* in dorsal view, showing the "arms."

Cyanogenesis, the production of hydrogen cyanide (HCN), has evolved independently in a number of arthropods, including centipedes (see Chapter 7), millipedes (see Chapter 9), beetles, hemipterans, and lepidopterans. Hydrogen cyanide is highly toxic, and its production by these animals is generally considered to be defensive.

Among the Lepidoptera, cyanogenesis has been shown to occur in certain moths of the family Zygaenidae, as well as in some butterflies of the family Nymphalidae. In cyanogenic Zygaenidae, hy-

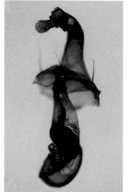
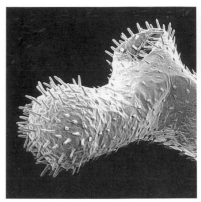

An enlarged view of the arm of *Calindoea trifascialis* (left), beside an arm that has been excised from a larva to show the large saclike gland that opens at the base of the arm (middle). The tip of the arm (right) shows the spines that presumably retain the secretory fluid after glandular discharge.

drogen cyanide production occurs in all developmental stages, from eggs to adult. Most recently, cyanogenesis has been reported from a moth of the family Thyrididae, one of the lesser-known lepidopteran taxa, containing mostly tropical and subtropical species.

The thyridid *Calindoea trifascialis* produces hydrogen cyanide in the larval stage. The species has been studied in a national park in Dac Lac Province, Vietnam, by Canadian investigators, who observed that when the larva was disturbed by handling it gave off an odorous fluid that appeared to be emitted from the base of two unusual armlike structures that projected from the sides of the animal. Several of the investigators, somewhat recklessly, put the fluid to the test by applying it to the tongue, and found that it promptly induced a localized numbing that persisted for about a minute. Subsequent laboratory study revealed the presence of two glands in the larva, positioned beneath the integument just behind the arms, and opening near the base of the arms by way of two narrow slits. The arms function as dabbing devices. They are terminally swollen and covered with flattened spines, which appear to retain the secretion after discharge. When the larva is disturbed, it flexes the body so that one arm is wiped against the disturbing agent. Field observation

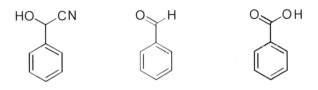

12. Mandelonitrile   14. Benzaldehyde   15. Benzoic acid

130. *(E,E)-α*-Farnesene   131. 3-Methylbutyl-3-methylbutanoate

has shown that through such action the larva effectively repelled ants. In fact, ants that were directly contacted by the arms were, in some cases, immediately immobilized.

Chemical analysis showed that the secretion of *C. trifascialis* contains mandelonitrile (12), a cyanogenic compound found also in some other hydrogen cyanide–producing arthropods. Mandelonitrile is the cyanohydrin of benzaldehyde—the adduct of hydrogen cyanide and benzaldehyde—and can be expected to yield free benzaldehyde when it dissociates in the course of cyanogenesis. Indeed, benzaldehyde (14) is typically found in the secretion of mandelonitrile-producing arthropods (see Chapters 7 and 9), and the compound was found in the secretion of *C. trifascialis* as well, together with oxidation products of benzaldehyde, such as benzoic acid (15). The *C. trifascialis* secretion also contained the isoprenoid (*E, E*)-α-farnesene (130) and the ester 3-methylbutyl-3-methylbutanoate (131).

The *C. trifascialis* secretion can be expected to act as a toxicant, on account of its contained hydrogen cyanide, and as a repellent (and irritant), by virtue of at least some of its other components. The numbing action of the secretion experienced by the investigators was almost certainly caused by the hydrogen cyanide, which could have been responsible also for the "knock-out" effect of the fluid on ants. Benzaldehyde must also contribute to the effectiveness of the secretion. The compound is known to be repellent to ants.

REFERENCES

Braekman J. C., D. Daloze, and J. M. Pasteels. 1982. Cyanogenic and other glucosides in a New Guinean bug *Leptocoris isolata*: possible precursors in its host-plant. *Biochemistry and Systematics* 10:355–364.

Darling, D. C. 2003. Morphology and behavior of the larvae of *Calindoea trifascialis* (Lepidoptera: Thyrididae), a chemically-defended retreat-building caterpillar from Vietnam. *Zootaxa* 225:1–16.

Darling, D. C., F. Schroeder, J. Meinwald, M. Eisner, and T. Eisner. 2001. Production of a cyanogenic secretion by a thyridid caterpillar. *Naturwissenschaften* 88:306–309.

Davis R. H., and A. Nahrstedt. 1985. Cyanogenesis in insects. In G. A. Kerkut and L. I. Gilbert, eds., *Comprehensive Insect Physiology and Pharmacology*. Oxford: Pergamon Press.

Franzl S., and C. M. Naumann. 1985. Cuticular cavities: storage chambers for cyanoglucoside-containing defensive secretions in larvae of a zygaenid moth. *Tissue and Cell* 17:267–278.

Franzl S., A. Nahrstedt, and C. M. Naumann. 1986. Evidence for site of biosynthesis and transport of the cyanoglucosides linamarin and lotaustralin in larvae of *Zygaena trifolii* (Insecta: Lepidoptera). *Journal of Insect Physiology* 32:705–709.

Nahrstedt, A. 1988. Cyanogenesis and the role of cyanogenic compounds in insects. In *Cyanide Compounds in Biology*, Ciba Foundation Symposium 140. Chichester: Wiley.

Nahrstedt, A., and R. H. Davis. 1986. (*R*)Mandelonitrile and prunasin, the sources of hydrogen cyanide in all stages of *Paropsis atomaria* (Coleoptera: Chysomelidae). *Zeitschrift für Naturforschung* 41c:928–934.

60

Class INSECTA

Order LEPIDOPTERA

Family Yponomeutidae

Ypsolopha dentella

The European honeysuckle leaf roller

An adult *Ypsolopha dentella.*

Ypsolopha dentella, a moth introduced into the United States from Europe, is a specialist, feeding on honeysuckle bushes. Well established in the northeastern United States, it makes its appearance as a larva early in the spring, feeding on the young leaves of the plant. The bizarre-looking adult is on the wing by mid-June.

Ypsolopha larvae are active primarily at night. They hide in the daytime, typically in a silken retreat, which they spin between leaves near the growing tips of branches. Their slender, spindle-shaped cocoons are affixed to the twigs of the plant and are well concealed. The adults are easy to detect but are quick to take flight and are not readily caught.

The larvae are most easily spotted by the telltale feeding injuries they inflict on leaves. They are also given away by their tiny black feeding pellets, which may accumulate outside their retreats unless blown away by the wind. The larvae are well camouflaged. Predominantly green, they match the color of the leaves they eat. A reddish streak on the back is of the precise hue often manifested by the petioles of honeysuckle leaves.

Aside from camouflage and the benefit of the retreat, the larvae have two additional defenses—they squirm and they "bungee-jump." Squirming, or wiggling, is an escape strategy that they share with many small caterpillars, but *Y. dentella* manifests this behavior in the extreme. Release a *Y. dentella* larva on a smooth surface and poke it gently with a toothpick, and it will propel itself over the surface by twisting and turning with such speed that it will literally vanish from view. The response may last only seconds, but that is long enough to allow the larva to distance itself from the scene. Even the youngest *Y. dentella* are programmed to wiggle when disturbed. Poke a larva that is hidden in its retreat and it will quickly wiggle away.

When a larva has wiggled its way to the edge of a leaf and can go no farther, it bungee-jumps, plunging into space, suspended by a slender silken thread that it squeezes from a small spigot that projects from just beneath the mouth. The spigot bears the openings of two large saclike glands that take up a considerable portion of the larva's body cavity and are bulging with the viscous liquid from which the silk is made. Bungee jumping does not require that the larva take time to anchor the silken thread to the edge of the leaf before jumping. The silken thread is fastened to the retreat itself and the larva needs only to let out a piece of silk long enough to cover the distance to the leaf edge. It is therefore connected to home base at all times and free to squirm away and take a bungee plunge at a moment's notice.

Ypsolopha dentella. Top left: A larva. Top middle: A pupal cocoon. Top right: A larva suspended from its silken "bungee" cord. Bottom left: A bundled bungee cord abandoned at the edge of the leaf after the larva ascended along the thread following a bungee drop. Bottom right: A larva wiggling away from an ant that has touched it.

When they plunge, the larvae drop some 10 to 20 centimeters or more, a considerable distance relative to a body length of 15 millimeters or less. In essence their plunging is another vanishing act, a trick that in combination with squirming enables the larvae to avoid trouble by leaving the scene. When the larvae plunge they drop virtually instantaneously. The rate at which they are able to convert liquid glandular precursor into silken strand is therefore quite remarkable.

The suspended larva, as a rule, does not remain hanging from its thread for long. Within minutes it begins the slow process of climbing up the thread to return to its retreat. To ascend, the larva pulls itself up along the cord, a short distance at a time, using its mouthparts to exert the pull and to tuck away the loops of slack reeled in as it proceeds upward. The process is elegant. The entire thread is stuffed into a holding space between the third pair of legs. When the larva completes its climb, it relinquishes the package, leaving it stuck to the leaf margin the moment it is back at home base. In so doing, the larva is essentially squandering a resource. Other silk producers are more conservation oriented. Spiders, for instance, when they take down their webs, eat the silk.

Bungee jumping is doubtless effective against many predators, including ants and jumping spiders, but it comes at a price. Aside from the cost of the silk itself, which the larva needs to conserve for eventual cocoon production, bungee jumping takes time. Although the larvae usually ascend within minutes, they sometimes remain suspended for over an hour. Potentially this wait could cut significantly into feeding time and therefore have a retarding effect on development. Whether it does depends on how often the larvae are attacked under normal circumstances, which has not been determined. But there is no question that predation pressure could be high at times upon *Y. dentella* larvae, particularly from ants.

One intriguing possibility is that bungee jumping protects *Y. dentella* larvae against predation by wasps. No data on *Y. dentella* itself are available, but other caterpillars have been shown to avoid wasp attacks by resorting to squirming behavior. In some of these caterpillars special integumental receptor hairs have been demonstrated that detect the air perturbations engendered by the wing vibrations of approaching wasps. Whether bungee jumping and squirming are activated by a similar sensory warning system in *Y. dentella* is not known.

Y. dentella belongs to the family Yponomeutidae, a relatively poorly known group of moths, including some 36 genera and nearly 1,000 species.

REFERENCES

Dougdale, J. S., N. P. Kristensen, G. S. Robinson and M. J. Scoble. 1999. The Yponomeutoidea. In N. P. Kristensen, ed., *Lepidoptera, Moths, and Butterflies, Handbook of Zoology*, vol. 1. New York: Walter de Gruyter.

Scoble, M. J. 1992. *The Lepidoptera: Form, Function, and Diversity.* Oxford: Oxford University Press.

Talekar, N. S., and A. M. Shelton. 1993. Biology, ecology, and management of the diamondback moth. *Annual Reviews of Entomology.* 38:275–301.

61

Class INSECTA
Order LEPIDOPTERA
Family Geometridae
Nemoria outina
A geometrid moth

The two larval forms of *Nemoria outina:* the leaf mimic and the twig mimic.

Caterpillars are anatomically peculiar in that they possess extra sets of legs. In addition to the three pairs of conventional thoracic legs situated just behind the head, most caterpillars have a set of short, fleshy, subsidiary legs, or prolegs, projecting from the midsection of the body and from the very rear. The prolegs are important because they enable the caterpillar to maintain a purchase along the full length of the body rather than just at the front end.

In one family of lepidopterans, the Geometridae, the caterpillars typically lack all prolegs except the two posteriormost pairs. They are therefore able to grab a hold only with the front and rear. Geometrid caterpillars are commonly known as loopers or inch-worms. When they move, they bring the front, then the rear forward stepwise, coiling the body into a loop each time the rear is advanced, thereby literally inching along.

· Geometrid caterpillars are masters of deception. They have capitalized on one form of defense—escaping detection—and they have taken this strategy to fascinating extremes.

Nemoria outina is a geometrid whose caterpillar lives exclusively on the rosemary bush *Ceratiola ericoides,* which grows in central Florida and bears no relationship to the garden herb of that name. The larva is remarkable in that it occurs in two forms that are drastically different in appearance. One form, the summer form, is green and of a shade and shape that enables it to blend in perfectly with the leaves of the plant. The other form, the winter form, is gray and rugose and precisely imitative of the twigs to which it clings when at rest. There is seasonal overlap of the two forms, so that they may at times both occur on the bush.

Another species of *Nemoria, N. arizonaria,* pulls a similar trick. Caterpillars of the spring brood of the species develop into mimics of the catkins of the oak tree upon which they feed. Caterpillars of the summer brood emerge after the catkins have fallen and they develop instead into mimics of twigs of the tree. Whether the caterpillars develop into one form or the other appears to depend on the concentration of phenolic compounds they ingest. Caterpillars feeding on catkins, which are low in phenolics, develop into catkin mimics, while those feeding on leaves, which are rich in phenolics, develop into twig mimics.

The duality of form illustrated by *Nemoria* is not common in geometrid caterpillars, but the imitation of plant parts per se is widespread. Many geometrid caterpillars have a twiglike appearance and they may behave in ways that enhance the imitation. Thus

A twig-mimicking larva of an unidentified geometrid moth. Note the imitation leaf scar on its body (enlarged, bottom right) and the resemblance of the front end (enlarged, top right) to the growing tip of a branch.

if they are on a twig when the plant is shaken, they sometimes straighten out and project rigidly at an angle from the twig, in imitation of side branches. Such larvae may also bear detailed imitations of leaf scars, and the front end is sometimes shaped very much like the growing tip of a branch.

Quite remarkable are the geometrid moths of the genus *Synchlora*. The larvae of these moths live on flowers, mostly on Asteraceae, and they disguise themselves as flowers. They cut pieces of petal from the flowers and fasten them to their bodies with strands of silk that they secrete from special glands. They have tubercles on their back with projecting spines, to which they tie the petal pieces. When the larvae pupate, they construct silken cocoons, taking care to weave the remnants of their floral cover into the fabric of the enclosure.

There is evidence also that some geometrids are chemically pro-

Left: A flower of *Bidens pilosa* (Asteraceae) bearing (on the left) a larva of an unidentified species of *Synchlora* that has cloaked itself with pieces of petals of the flower and is virtually invisible as a result. Right: The same larva, exposed by removal of its petal cover.

tected, both in the larval and the adult stages. The defensive chemicals have been isolated and characterized in some cases, and shown in some instances to be derived from the foodplants.

The Geometridae are a successful group of moths. They include some 1,500 genera and 20,000 species.

REFERENCES

Greene, E. 1989. A diet-induced developmental polymorphism in a caterpillar. *Science* 243:643–646.

Heitzman, R. L. 1982. Descriptions of the mature larva and pupa of *Hypomecis umbrosaria* (Lepidoptera, Geometridae). *Proceedings of the Entomological Society of Washington* 84:111–116.

Nishida, R., M. Rothschild, and R. Mummery. 1994. A cyanoglucoside, sarmentosin, from the magpie moth, *Abraxas grossulariata* (Geometridae: Lepidoptera). *Phytochemistry* 36:37–38.

Nishida, R., H. Fukami, R. Iriye, and Z. Kumazawa. 1990. Accumulation of highly toxic ericaceous diterpenoids by the geometrid moth *Arichanna gaschkevitichii*. *Agricultural and Biological Chemistry* 54:2347–2352.

Yasui, H. 2001. Sequestration of host plant-derived compounds by geometrid moth, *Milionia basalis,* toxic to a predatory stink bug, *Eocanthecona furcellata. Journal of Chemical Ecology* 27:1345–1353.

62

Class **INSECTA**

Order **LEPIDOPTERA**

Family Arctiidae

Utetheisa ornatrix

The rattlebox moth

A spider *(Argiope florida)* inspecting a *Utetheisa ornatrix* before setting the moth free.

The moth *Utetheisa ornatrix* reacts in an unusual way when it flies into a spider web. Instead of attempting to flutter loose, it simply folds its wings and becomes quiescent. The spider, reacting as it typically does to the impact of an insect in its web, darts toward the moth and inspects it. It probes it with its palps and front legs, some-

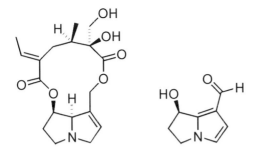

132. Monocrotaline   **133. Hydroxydanaidal**

times for half a minute or so, but then proceeds to do an odd thing. It cuts the moth free. *U. ornatrix* is evidently unacceptable to orb-weaving spiders, and as we know from experiments, the unacceptability extends to include other spiders as well. Wolf spiders, for instance, also reject the moth.

As a larva, *U. ornatrix* feeds on leguminous plants of the genus *Crotalaria.* These plants have long been known to be poisonous, by virtue of certain alkaloids—called pyrrolizidine alkaloids, or PAs (for example, monocrotaline, **132**)—they contain. *U. ornatrix* larvae are unaffected by PAs and store the toxins in their bodies, retaining them through metamorphosis into the adult stage. *U. ornatrix* does not require PAs for normal development. The moth can be reared in the laboratory on a pinto bean–based diet totally free of PAs. Adults raised on that diet are normal in appearance, but, lacking PAs, are palatable to spiders.

As a larva, *U. ornatrix* is already protected by PAs. Wolf spiders, for instance, reject *U. ornatrix* caterpillars that are reared on their normal PA-containing foodplant, but eat caterpillars that are raised on the pinto bean–based diet.

When attacked, the adult *U. ornatrix* often emits a froth from the neck region, a bubbling mass made up of blood and air. In moths reared on their normal foodplant, the froth contains PAs, and is assumed to contribute to the moth's defense.

Female *U. ornatrix* transmit some of their PAs to the eggs. These are protected as a result, from enemies as diverse as ants, ladybird beetles, green lacewing larvae, and parasitic wasps. Eggs laid by fe-

Utetheisa ornatrix frothing.

males raised on the pinto bean–based diet are PA-free and vulnerable to attack.

A female *U. ornatrix* lays hundreds of eggs, and she may not possess enough PA to protect them all. But she is able to obtain supplementary PA by mating. A male transmits PA to the female with the sperm package, a sizable structure that also contains nutrients and may amount to over 10% of the male's body mass. Females make it a point to mate repeatedly. From counts of remnants of the sperm packages recovered from the female's sperm pouch (the bursa) it was determined that females in nature mate on average with 11 males. Champions among them take more than 20 partners. The benefits from such behavior add up. The more often the female mates, the more PA she is able to accrue.

Before taking a male as a partner, the female quite literally demands proof from him that he holds a substantial amount of PA in store for her. The male provides such proof with a pheromone, a chemical called hydroxydanaidal (or HD) (**133**), which he derives from his systemic PA in amounts proportional to his PA content. He dispenses the pheromone with a pair of brushes that he ordi-

Utetheisa ornatrix. Top left: A larva emerging from the egg. Bottom left: A nearly fully grown larva inside a pod of *Crotalaria spectabilis,* feeding on seeds. Top right: A mating pair. Bottom right: The male's pheromone-bearing brushes.

narily keeps tucked away but everts from the abdominal tip during precopulatory interplay with the female. Males differ in their PA content, probably because they differ in their ability to sequester PA as larvae. The PA is concentrated in the seeds of the *Crotalaria* plants, and males may not all have equal access to these seeds. By favoring males with high HD content, females choose partners that have excelled in the larval competition for seeds. Such males, more richly HD-scented and PA-endowed, are also the ones able to bestow the largest PA gifts, to the females' great advantage. Males richest in PA are also the largest, which in itself is of consequence, because size is a heritable trait in *U. ornatrix*. By favoring males with high HD content, therefore, females ensure not only that they receive larger PA gifts (a phenotypic benefit) but that they engender larger offspring (a genetic benefit). Having large offspring pays off on two counts: larger sons are at an advantage in courtship, and larger daughters are more fecund.

The female herself also benefits from the PA she receives from males. Experiments have shown that she is able to allocate that PA to virtually all body parts and that she derives defensive advantage as a result.

Evidence also indicates that females do not give all accrued sperm an equal chance. The females appear to favor sperm received in larger sperm packages, which also happen to be those delivered by the larger males, bearing the better genes. Although selective in their use of sperm, the females are not choosy with regard to the utilization of their PA gifts. They accept these from all partners, and put them to defensive use irrespective of their source.

Parental bestowal of defensive chemicals upon eggs has been demonstrated for a diversity of insects. Interestingly, in some of these cases, the donating parent is the father, who, as in *U. ornatrix,* transmits the chemical to the mother for eventual incorporation into the eggs. Also remarkable is that the protective substances, in a number of species (including other arctiid moths and danaine butterflies), have been shown to be PAs.

REFERENCES

Bezzerides, A., and T. Eisner. 2002. Apportionment of nuptial alkaloidal gifts by a multiply-mated female moth (*Utetheisa ornatrix*): eggs individually receive alkaloid from more than one male source. *Chemoecology* 12:213–218.

Bezzerides, A., T.-H. Yong, J. Bezzerides, J. Husseini, J. Ladau, M. Eisner, and T. Eisner. 2004. Plant-derived pyrrolizidine alkaloid protects eggs of a moth *(Utetheisa ornatrix)* against a parasitoid wasp

(*Trichogramma ostriniae*). *Proceedings of the National Academy of Sciences USA* 101:9029–9032.

Conner, W. E., R. Boada, F. C. Schroeder, A. González, J. Meinwald, and T. Eisner. 2000. Chemical defense: bestowal of a nuptial alkaloidal garment by a male moth upon its mate. *Proceedings of the National Academy of Sciences USA* 97:14406–14411.

Eisner, T., and J. Meinwald. 2003. Alkaloid-derived pheromones and sexual selection in Lepidoptera. In G. J. Blomquist and R. G. Vogt, eds., *Insect Pheromone Biochemistry and Molecular Biology.* London: Elsevier Academic Press.

Eisner, T., C. Rossini, A. González, V. K. Iyengar, M. V. S. Siegler, and S. R. Smedley. 2002. Paternal investment in egg defense. In M. Hilker and T. Meiners, eds., *Chemoecology of Insect Eggs and Egg Deposition.* Berlin: Blackwell Publishing.

Hartmann, T., and L. Witte. 1995. Chemistry, biology, and chemoecology of the pyrrolizidine alkaloids. In S. W. Pelletier, ed., *Alkaloids: Chemical and Biological Properties.* Oxford: Pergamon Press.

Iyengar, V. K., H. K. Reeve, and T. Eisner. 2002. Paternal inheritance of a female moth's mating preference. *Nature* 419:830–832.

Nishida, R. 2002. Sequestration of defensive substances from plants by Lepidoptera. *Annual Review of Entomology* 47:57–92.

Rossini, C., A. Bezzerides, A. González, M. Eisner, and T. Eisner. 2004. Chemical defense: incorporation of diet-derived pyrrolizidine alkaloid into the integumental scales of a moth *(Utetheisa ornatrix). Chemoecology* 13:199–205.

63

Class INSECTA

Order LEPIDOPTERA

Family Saturniidae

Automeris io

The io moth

An *Automeris io* larva.

Fleeting contact with the integument of an insect is usually of no consequence. The exceptions, however, can be dramatic, especially if they involve brushings with urticating caterpillars. The term urticating, from the Latin *urticare,* to sting, is appropriate to a number of lepidopteran larvae known more familiarly as stinging hair caterpillars. There are many such species in the tropics, where lepidopteran diversity is at a peak. In North America the most familiar urticating caterpillar is that of the io moth, *Automeris io,* a member of the giant silkworm moth family, the Saturniidae, which includes some of the largest, most beautiful moths in the world. *A. io* is best known as an adult because it is attracted to light and sometimes comes to settle on window screens, where it is readily apparent because of its large size (its wing span is 5–8 centimeters). When disturbed while at rest, as by an attacking predator, the moth displays a remarkable startle response, in which it spreads the forewings, abruptly exposing a pair of huge, white-centered black eyespots on the upper surface of its hindwings. The behavior is surely a deterrent to predators, especially birds. Captive birds, while engaged in eating at a feeding station, have been shown to be frightened off when suddenly presented with paired eye images.

By contrast, the pale greenish-yellow caterpillar of *A. io* is often inconspicuous on its foodplant. The larva feeds on a broad diversity of plants, including oaks, willows, maples, and birches. Anybody who brushes against an io moth larva with bare skin will experience instantaneous localized pain, followed by itching and inflammatory

A close-up view of the poisonous spines of the *Automeris io* larva.

swelling. The cause of the discomfort is the urticating hairs—or, rather, spines, for they are more robust than hairs—that project from the caterpillar's back. Sharply pointed, these spines readily penetrate human skin. Each has an associated secretory cell at the base, and is replete with venom. The spines rupture as they pierce the skin, thus releasing their toxic contents. Given their effects on humans, there can be no doubt that the spines are protective against natural enemies.

Little effort has been invested in the study of the chemical composition of the venom, partly because human encounters with the larva are relatively rare and because the effects of the venom are not life threatening. One is tempted to argue, by analogy with what is known from other urticating caterpillars, that the venom of *A. io* contains pain-inducing factors such as histamine and serotonin, but the possibility remains that the fluid contains specialized toxins of its own.

Other saturniid caterpillars can inflict more serious injury to humans. Notable are the South American members of the genus *Lonomia.* Contact with these caterpillars induces an immediate localized burning sensation, followed sometimes by delayed intracerebral and intraperitoneal hemorrhaging. In extreme cases the sequel can be renal failure and death. The venom of *Lonomia achelous* larvae, one of the most dangerous species, contains proteinaceous prothrombin activators as well as a variety of fibrinolytic enzymes, which contribute to the hemorrhaging by degrading fibrinogen. Many other South American species of stinging hair caterpillars are cause for concern, including *Automeris liberia,* whose venom provokes acute pain and necrosis of the skin. The composition of its venom is unknown.

Urticating caterpillars worldwide include species from families other than the Saturniidae, notably the Limacodidae, Arctiidae, Megalopygidae, and Lymantriidae. The Lymantriidae include the brown-tailed moths of the genus *Euproctis* (of which *Euproctis chrysorrhoea* occurs in North America), the venom of which has been shown to contain histamine, as well as proteolytic enzymes that elicit pain by promoting kinin release from plasma. Venom components from other caterpillars are known only imprecisely, and lit-

Top: *Automeris io* at rest. Bottom: *A. io* with the hindwings exposed, following disturbance. The abrupt presentation of the "eyes," startling to humans, undoubtedly startles predators as well.

tle has been learned of the action of these venoms on predators. Venoms of other invertebrates such as spiders and cone snails have been examined comprehensively in recent years. It is surprising that caterpillar venoms have been largely excluded from these studies. There is no reason why caterpillar venoms should not turn out to contain novel toxins of interest.

REFERENCES

Arocha-Pinango, C. L., E. Marval, and B. Guerrero. 2000. *Lonomia* genus caterpillar toxins: biochemical aspects. *Biochimie* 82:937–942.

Blest, A. D. 1957. The function of eyespot patterns in the Lepidoptera. *Behaviour* 11:209–255.

Delgado Quiroz, A. 1978. Venoms of Lepidoptera. In S. Bettini, ed., *Handbook of Experimental Pharmacology,* vol. 48, *Arthropod Venoms.* Berlin: Springer-Verlag.

Jones, D. L., and J. H. Miller. 1959. Pathology of the dermatitis produced by the urticating caterpillar *Automeris io. American Medical Association Archives of Dermatology* 79:81–85.

Kawamoto, F., and N. Kumada. 1984. Biology and venoms of Lepidoptera. In A. T. Tu, ed., *Handbook of Natural Toxins,* vol. 2, *Insect Poisons, Allergens, and Other Invertebrate Venoms.* New York: Marcel Dekker.

Lamdin, J. M., D. E. Howell, K. M. Kocan, D. R. Murphey, D. C. Arnold, A. W. Fenton, G. V. Odell, and C. L. Ownby. 2000. The venomous hair structure, venom and life cycle of *Lagoa crispata,* a puss caterpillar of Oklahoma. *Toxicon* 38:1163–1189.

Pesce, H., and A. Delgado. 1971. Poisoning from adult moths and caterpillars. In W. Bücherl and E. E. Buckley, eds., *Venomous Animals and Their Venoms,* vol. 3, *Venomous Invertebrates.* New York: Academic Press.

64

Class INSECTA
Order LEPIDOPTERA
Family Papilionidae
Eurytides marcellus
The zebra swallowtail butterfly

A pupa of *Eurytides marcellus.*

The caterpillars of butterflies of the family Papilionidae, the swallowtails, have a glandular defensive organ, the osmeterium, which they keep tucked away behind the head and evert when disturbed. The osmeterium is a two-pronged structure, which upon extrusion gives the larva a horned appearance. The two prongs, when involuted, are kept full of secretion. When turned inside out upon eversion, they are coated with the fluid. The larvae combine osmeterial eversion with postural adjustments of the body. When touched in the rear, for instance, they arch up and flex the front end backward, wiping the osmeterial prongs against the "enemy." The secretion is potently active even in vapor form. Predators are sometimes repelled before actually being touched by the organ. Both invertebrate and vertebrate predators have been shown to be deterred by the osmeterium.

In the caterpillar of the zebra swallowtail butterfly, *Eurytides marcellus,* the osmeterial secretion contains two potently odorous compounds, isobutyric acid (**134**) and 2-methylbutyric acid (**135**). The osmeterial prongs are pale and translucent, and as is generally the case with swallowtail larvae, are everted only in response to contact stimulation of the body. Also shared with other swallowtail larvae is *E. marcellus*'s habit of regurgitating gut fluid at the same time that the osmeterium is everted. The emitted fluid may spread from the mouth onto adjacent areas, including even the osmeterial prongs, and if so the predator will then be contacted by a mixture of enteric and glandular fluids (see Chapter 39). The defensive effectiveness of the mixture may well be enhanced by the presence in the oral effluent of repellent compounds from the foodplant.

E. marcellus derives protection further from the mimetic resemblance of its pupa to the leaves of its foodplant, pawpaw (*Asimina* species). The larva pupates amidst these leaves and may then be very difficult to detect. The pupa is the same shade of green as the leaves and bears a surface pattern of lines and ridges closely imitative of the leaves' venational network.

A number of other swallowtail butterflies also produce isobutyric acid and 2-methylbutyric acid as part of their osmeterial secretion. Among these are several of the most familiar North American swallowtails, including *Papilio troilus* (the spicebush swallowtail), *Papilio glaucus* (the tiger swallowtail), *Papilio palamedes* (the palamedes swallowtail), *Papilio polyxenes* (the black swallowtail), and *Papilio cresphontes* (the giant swallowtail). But there are species that produce entirely different osmeterial compounds. *Battus polydamas*

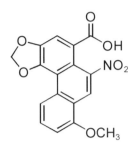

134. Isobutyric acid   **135.** 2-Methylbutyric acid   **136.** β-Selinene

137. Selin-11-en-4α-ol   **138.** Aristolochic acid

(the polydamas swallowtail), for instance, produces two sesquiter-penes, β-selinene (**136**) and selin-11-en-4α–ol (**137**). Other compounds, including a diversity of simple isoprenoids, have been characterized from the osmeterial fluid of several other swallowtail species. There is further evidence that the composition of the osmeterial secretion may vary both qualitatively and quantitatively from individual to individual in certain swallowtail species, or as a function of the larval age. There may even be a difference in the output of the two prongs of the same osmeterium.

Work with radio-labeled precursors has shown that at least insofar as certain of the terpenoid components of osmeterial secretion are concerned, papilionid larvae come upon these substances by endogenous synthesis rather than by incorporation from the diet.

As effective as the osmeteria appear to be, they do not protect papilionid larvae against all enemies. The pentatomid bug *Podisus maculiventris,* for example, is able to feed on *P. polyxenes* larvae without inducing osmeterial eversion. It pierces the larva with its beak and sucks out its body fluids. The bug is actually repelled by the osmeterial secretion but by being able somehow to keep the larva

Top: Larvae of *Eurytides marcellus* (left) and *Battus philenor* (right) everting the osmeterium in response to disturbance. The forceps used to stimulate the larva are visible in the photo on the left. Bottom: A larva of *Papilio troilus*. Note that although the larva is viewed from two different directions its "eyes" appear to stare directly back at the viewer.

from extruding its defensive organ it avoids exposure to the glandular fluid.

It is not uncommon for swallowtail larvae to have defenses in addition to the osmeterium. *P. troilus* larvae, for example, have a pair of highly conspicuous eye images near their front end, which give the uncanny impression of "staring in defiance," and may for that reason intimidate some predators (see Chapter 63). Young *P. troilus* larvae lack the eye images. Instead they bear an irregular pattern of black and white markings, as do bird droppings, which they distinctly resemble. They are not unique among swallowtail larvae in being thus imitative. Young larvae of *P. cresphontes* and of *P. polyxenes* masquerade as avian droppings as well. The white markings in *P. polyxenes* larvae have been shown to be formed by accumulated uric

A young larva of *Paplio troilus* (left) next to a bird dropping that it imitates.

acid crystals. Although uric acid in insects has traditionally been viewed as an excretory product, it can also act, when stored systemically, as a powerful antioxidant. Thus it is possible that in papilionid larvae it serves for protection against oxidative stress generated by toxins ingested with the foodplant, while at the same time providing the animals with the white pigment they require to generate their visual disguise. The strategy of imitating avian feces is not limited to swallowtail larvae. There are moths that as adults show the mottled black and white patterning typical of bird droppings, and such moths may have the habit of resting during the daytime on the upper surface of leaves (where bird feces are likely to be found), rather than on the underside, as is typical for moths.

Swallowtail butterflies also benefit as adults from their rapid, erratic flight. As any butterfly collector knows, it takes skill and a net of considerable size to capture swallowtails when they are on the wing. Some swallowtails are also protected by being distasteful as adults. Members of the genus *Battus,* such as the North American *B. philenor* (the pipevine swallowtail) and *B. polydamas,* possess aristolochic acids (for example, **138**), powerful feeding deterrents that they sequester from their larval foodplants (species of *Aristolochiaceae*). Aristolochic acids appear also to protect the larvae of *Battus,* as for instance against parasitoid ichneumonid wasps.

REFERENCES

Berenbaum, M. R., B. Moreno, and E. Green. 1992. Soldier bug predation on swallowtail caterpillars (Lepidoptera: Papilionidae): circumvention of defensive chemistry. *Journal of Insect Behavior* 5:547–553.

Brower, L. P. 1984. Chemical defense in butterflies. In R. I. Vane-Wright and P. R. Ackery, eds., *The Biology of Butterflies.* London: Academic Press.

Burger, B. V., Z. Munro, M. Roth, H. S. C. Spies, V. Truter, H. Geertsema, and A. Habich. 1985. Constituents of osmeterial secretion of pre-final instar larvae of citrus swallowtail *Papilio demodocus* (Lepidoptera, Papilionidae). *Journal of Chemical Ecology* 11:1093–1114.

Eisner, T., and Y. C. Meinwald. 1965. Defensive secretion of a caterpillar (*Papilio*). *Science* 150:1733–1735.

Eisner, T., A. F. Kluge, M. I. Ikeda, Y. C. Meinwald, and J. Meinwald. 1971. Sesquiterpenes in the osmeterial secretion of a papilionid butterfly, *Battus polydamas. Journal of Insect Physiology* 17:245–250.

Eisner, T., T. E. Pliske, M. Ikeda, D. F. Owen, L. Vázquez, H. Pérez, J. G. Franclemont, and J. Meinwald. 1970. Defense mechanisms of arthropods. XXVII. Osmeterial secretions of papilionid caterpillars *(Baronia, Papilio, Eurytides). Annals of the Entomological Society of America* 63:914–915.

Honda, K. 1981. Larval osmeterial secretions of the swallowtails *Papilio. Journal of Chemical Ecology* 7:1089–1114.

———Defensive potential of components of the larval osmeterial secretion of papilionid butterflies against ants. *Physiological Entomology* 8:173–180.

——— GC-MS and carbon-13 NMR studies on the biosynthesis of terpenoid defensive secretions by the larvae of papilionid butterflies *Luehdorfia* and *Papilio. Insect Biochemistry* 20:245–250.

Sime, K. 2002. Chemical defence of *Battus philenor* larvae against attack by the parasitoid *Trogus pennator. Ecological Entomology* 27:337–345.

Stamp, N. E. 1986. Physical constraints of defense and response to inver-

tebrate predators by pipevine caterpillars *Battus philenor*
Papilionidae. *Journal of the Lepidopterists' Society* 40:191–205.

Timmerman, S., and M. R. Berenbaum. 1999. Uric acid deposition in
larval integument of black swallowtails and speculation on its possible functions. *Journal of the Lepidopterists' Society* 53:104–107.

65

Class INSECTA
Order LEPIDOPTERA
Family Pieridae
Pieris rapae
The cabbage butterfly

Pieris rapae at rest.

The propensity of humans to travel and to transport goods by truck, train, and plane has created and continues to create an immense opportunity for organisms to spread beyond their natural range. Many insects have been introduced to new regions in this way, with consequences that have on occasion proven disastrous, as when such species join the ranks of local agricultural pests or become vectors of disease. Species differ with regard to the readiness with which they adapt to new environments. The honey bee, for example, which has spread worldwide with human help, proved capable of fitting into virtually any new landscape because of the ease with which it could shift to the exploitation of new flowers. The honey bee was doubtless also helped in its spread by possession of the stinger, which must have provided it with the capacity to cope with new predation pressures.

Possession of defenses may in fact be one of the major determinants of a species' ability to colonize new domains. An interesting example is the cabbage butterfly, *Pieris rapae,* one of the most familiar lepidopterans and a common pest. *P. rapae* is currently widely distributed. A native of Eurasia and North Africa, it was introduced into Canada around 1860, from where it spread over most of North America and eventually to Bermuda, Australia, Hawaii, and other Pacific Islands. The spread of *P. rapae* was greatly aided by the fact that its foodplants, including cabbage, Brussels sprouts, and cauliflower, were cultivated the world over. But the insect's adaptability to new regions was doubtless also a consequence of its possessing effective defenses.

There is evidence, for example, that *P. rapae* is protected as an adult. Some of the data stem from an investigation carried out in the early 1930s on the island of Martha's Vineyard in Massachusetts by an investigator, Frank Morton Jones, who was interested in the acceptability of insects to birds. To test for this, he laid out freshly killed insects on trays, and presented these trays to birds outdoors, while he kept precise track of which insects were eaten and which were spurned. *P. rapae* received a very low acceptability rating. Placed on a tray with an assortment of moths and other insects, the *P. rapae* were almost all left behind (most bore "mutilations," indicating that they had been "picked up by the birds and

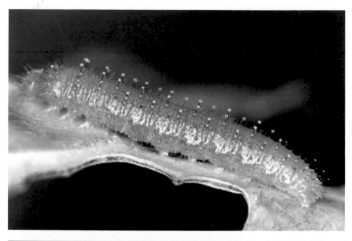

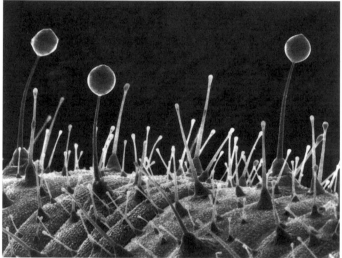

Top: A *Pieris rapae* larva. Bottom: A close-up view of the larva showing the secretion-bearing glandular hairs.

then dropped"), while the other offerings were mostly eaten. The visiting birds were towhees, blue jays, and robins. More recent evidence indicates that the unacceptability of adult *P. rapae* has a chemical basis, but neither the nature of the compounds responsible nor whether they are of exogenous origin or produced by the insect itself has been established.

P. rapae is also protected as a larva. Its weaponry in the caterpillar

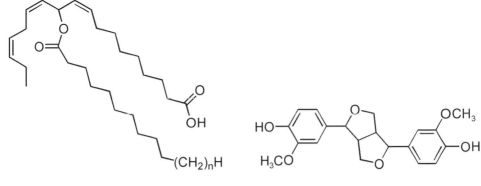

139. A Mayolene
140. Pinoresinol

stage takes the form of glandular hairs, neatly arranged in rows along the back and flanks of the animal, and bearing at their tips droplets of a clear oily fluid. The liquid originates in glandular cells at the base of the hairs and is conveyed through the hollow of the hairs to the tips.

Experiments showed that ants react promptly to contact with the hairs. The secretory droplets spread readily onto the surface of the ants, causing them to engage in protracted cleaning activities and keeping them from inflicting injury.

Chemical analysis showed that the secretion consists primarily of a series of unsaturated lipids, which because of their anomalous structure were given a special name: mayolenes (for example, **139**). Tests with synthetic mayolenes, applied to food items ordinarily taken by ants, proved that the chemicals render these items unacceptable, which leaves no doubt that the mayolenes are at least partly responsible for the defensive potency of the secretion. Further analysis showed the larval secretion also to contain low concentrations of pinoresinol (**140**), a member of a large class of plant compounds called lignans. Pinoresinol is most probably of dietary origin in *P. rapae,* and it contributes to the effectiveness of the secretion.

Glandular hairs have received far less attention in insects than in plants. They are rarer in insects, but no less worthy of study. Indeed, the two types of compounds most recently characterized from the secretion of insectan glandular hairs—the mayolenes from *P.*

rapae and the polyazamacrolides from *Epilachna* beetles (see Chapter 46)—are both of considerable chemical interest.

The secretory products of glandular hairs must conform to certain specifications. Since such products are deployed externally, they need to be chemically stable and relatively nonvolatile. They are therefore likely to differ chemically from the typical low-molecular-weight, volatile substances that insects discharge when attacked and store in internal sacs. For this reason alone, that they stand a chance of being anomalous, insectan glandular hair products are worth characterizing. An easy first step could be the investigation of pierid butterflies related to *P. rapae*. Some of these also have glandular hairs, but they have so far not been studied chemically.

REFERENCES

Brower, L. P. 1984. Chemical defense in butterflies. In R. I. Vane-Wright and P. R. Ackery, eds., *The Biology of Butterflies*. London: Academic Press.

Jones, F. M. 1932. Insect coloration and the selective acceptability of insects to birds. *Transactions of the Royal Entomological Society London* 80:345–385.

Müller, C., N. Agerbirk, and C. E. Olsen. 2003. Lack of sequestration of host plant glucosinolates in *Pieris rapae* and *P. brassicae. Chemoecology* 13:47–54.

Opler, P. A., and G. O. Krizek. 1984. *Butterflies East of the Great Plains*. Baltimore: Johns Hopkins University Press.

Schroeder, F. C. Unpublished data on *Pieris rapae*.

Smedley, S. R., F. C. Schroeder, D. B. Weibel, J. Meinwald, K. A. LaFleur, J. A. Renwick, R. Rutowski, and T. Eisner. 2002. Mayolenes: labile defensive lipids from the glandular hairs of a caterpillar *(Pieris rapae)*. *Proceedings of the National Academy of Sciences USA* 99:6822–6827.

66

Class **INSECTA**
Order **LEPIDOPTERA**
Family Nymphalidae
Danaus plexippus
The monarch butterfly

The monarch butterfly, *Danaus plexippus.*

The monarch butterfly, *Danaus plexippus,* is one of the few insects that can be said to have achieved true popularity. Widely admired for its migratory habits, the monarch is also well known, both to biologists and science buffs, for being distasteful and for serving as the model for the viceroy butterfly, *Limenitis archippus,* a look-alike of the monarch that does not share the latter's unpalatability.

The monarch derives its distastefulness from certain steroidal

58. Calotropin

substances called cardenolides, or cardiac glycosides (for instance calotropin, **58**; see also Chapter 25), which it acquires from the milkweed plants (family Asclepiadaceae) on which it lives and feeds as a caterpillar.

The monarch butterfly holds keystone status within chemical ecology, and rightly so. It is typically *the* species that is invoked to exemplify the phenomenon of defensive metabolite acquisition by insects, a phenomenon that is also often cited as illustrative of coevolutionary interaction. Secondary metabolites in plants, it is argued, are largely defensive and have been responsible, over evolutionary time, for narrowing the spectrum of enemies of specific plants or groups of plants. Some insects in turn, again over evolutionary time, have breached the defenses of these plants by becoming tolerant of their metabolites, and once tolerant, have evolved the capacity to use the plant's metabolites for defensive purposes of their own. Hence the monarch.

Unpalatability appears to be a common trait among the 200 species of butterflies worldwide, mostly from Asia and Africa, making up the nymphalid subfamily Danainae, to which the monarch belongs. In North America, for example, there is the queen butterfly, *Danaus gilippus,* which also derives distastefulness from feeding on milkweed plants, and in Africa there are, among others, the species of the genus *Amauris.* Some of these species make use of defensive substances beyond those they acquire from the larval foodplants. The queen butterfly, for instance, protects its eggs with pyrrolizidine alkaloids (see also Chapter 62) that the adult males acquire by im-

Danaus plexippus (left) next to its mimic, the viceroy butterfly, *Limenitis archippus* (right).

bibing the excrescences of certain plants and transmit to the female with the sperm package.

The cardenolides acquired by the monarch butterfly and other species of *Danaus* from milkweed plants are not only intrinsically noxious by virtue of bitterness, but emetic. If a blue jay that is unfamiliar with monarchs eats one, there is a good chance the bird will vomit minutes later. Moreover, it will probably learn from the experience. By associating the taste signals present in the food as it is being regurgitated with the memory of the dietary item that bore the signals, it can come to discriminate against monarchs, first on the basis of taste and then, after associating image with flavor, on the basis of appearance alone. Having thus become "monarch-shy," the jay would then be expected to discriminate against mimics of the monarch as well. Hence the existence of the viceroy butterfly.

The literature on the palatability of danaine butterflies generally, and of the monarch in particular, is extensive, and of great interest. The field of insect-plant relationships is still expanding on many fronts, as we learn more and more about ways in which insects put acquired plant materials to use (see Chapters 23, 25, 64, and 67).

Although utilization of plant material for defensive purposes has now been documented for a large number of animals, the utilization by animals of defensive products from other animals has only relatively infrequently been noted. The European hedgehog, for instance, which regularly consumes toads and is itself insensitive to the noxious exocrines from the skin glands of these animals, anoints

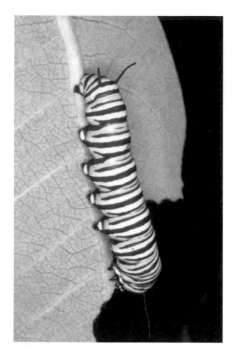

A larva of *Danaus plexippus* on one of its
foodplants, the milkweed *Asclepias syriaca.*

itself with these products. It takes pieces of toad skin in the mouth,
and after chewing them vigorously and mixing them with copious
quantities of saliva, spits the concoction onto the spines on its back.
The treatment apparently increases the defensive effectiveness of the
spines (for the use of frog toxins by indigenous humans, see Chapter
68). Similarly, capuchin monkeys in Venezuela anoint themselves
with benzoquinone-producing millipedes. They rub the millipedes
into their fur, thereby causing the animals to discharge their quinones (see Chapter 8). Wetted with secretion, the monkeys are presumed to lose their attractancy to mosquitoes, which are known vectors of diseases.

No animals are as versatile as humans in putting acquired substances to use. There are literally hundreds of natural products that
we appropriate for one purpose or another from plants, microbes, or
animals, or that we manufacture in imitation or near imitation of
compounds originally derived from the biotic world. Medicinals
that we have come upon by exploration of nature are among our

most important protective agents. In a way, in our strategy of appropriating defensive chemicals from nature, we are mere imitators of the likes of the monarch butterfly and other insects that pioneered the technique well before we ourselves became a presence on Earth.

REFERENCES

Brodie, E. D. 1977. Hedgehogs use toad venom in their own defense. *Nature* 268:623–628.

Brower, J. V. Z. 1958. Experimental studies of mimicry in some North American butterflies. I. The monarch *Danaus plexippus* and viceroy *Limenitis archippus. Evolution* 12:32–47.

Brower, L. P. 1984. Chemical defence in butterflies. In R. I. Vane-Wright and P. R. Ackery, eds., *The Biology of Butterflies.* London: Academic Press.

Dussourd, D. E., C. A. Harvis, J. Meinwald, and T. Eisner. 1989. Paternal allocation of sequestered plant pyrrolizidine alkaloid to eggs in the danaine butterfly, *Danaus gilippus. Experientia* 45:896–898.

Ehrlich, P. R., and P. H. Raven. 1965. Butterflies and plants: a study in coevolution. *Evolution* 18:586–608.

Nishida, R. 2002. Sequestration of defensive substances from plants by Lepidoptera. *Annual Review of Entomology* 47:57–92.

Rothschild, M. 1972. Secondary plant substances and warning coloration in insects. *Symposium of the Royal Entomological Society London* 6:59–83.

Rothschild, M., and D. Kellert. 1972. Reactions of various predators to insects storing heart poisons in their body tissues. *Journal of Entomology* A 46:103–110.

Swynnerton, C. F. M. 1915. Birds in relation to their prey: experiments on wood-hoopoes, small hornbills, and a babbler. *Journal of the South African Ornithological Union* 1915:22–108.

Valderrama, X., J. G. Robinson, A. B. Attygalle, and T. Eisner. 2000. Seasonal anointment with millipedes in a wild primate: a chemical defense against insects? *Journal of Chemical Ecology* 26:2781–2790.

Weldon, P. J., J. R. Aldrich, J. A. Klun, J. E. Oliver, and M. Debboun. 2003. Benzoquinones from millipedes deter mosquitoes and elicit self-anointing in capuchin monkeys (*Cebus* spp.). *Naturwissenschaften* 90:301–304.

67

Class INSECTA
Order HYMENOPTERA
Family Pergidae
Perga affinis
A pergine sawfly

Perga affinis larvae, aggregated on a eucalypt tree.

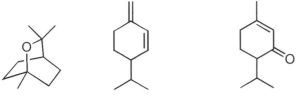

141. 1,8-Cineole   142. β-Phellandrene   143. Piperitone

Few insects eat the leaves of eucalypt trees (family Myrtaceae), and for good reason. Eucalypt leaves are rich in a highly odorous oil, laden with chemicals repellent to insects. The oil might well have contributed to the eucalypts' evolutionary success. In their native Australia, the eucalypts, or "gum trees," have diversified into hundreds of species.

Eucalypts produce the oil within the leaves themselves, where it is stored in small globular cavities present throughout the leaf parenchyma. Typically, the oil is composed of a mixture of isoprenoids [for example 1,8-cineole (141), β-phellandrene (142), and piperitone (143)], and diverse other volatiles, including aldehydes, alcohols, and phenols.

There can be no question that eucalypts derive protection from possession of the oil, but there are, as you might expect, insects that have evolved ways of getting around the eucalypts' defense. Interesting among them are the larvae of certain sawflies, whose specialty it is to feed on eucalypts. Sawflies are not flies but Hymenoptera, members of that large order of insects that includes also the bees, wasps, and ants. Sawflies are primitive hymenopterans. Those that feed on eucalypts, including species of the genera *Perga* and *Pseudoperga* (family Pergidae, subfamily Perginae), have the remarkable ability to separate the oil from the leaf tissue as they ingest the leaves, swallowing only the tissue and shunting the oil into a large diverticulum of the esophagus. How exactly the larvae separate the oil from the tissue is not entirely understood, although the mandibles in some species are equipped with a mesh of tiny papillae that may serve the purpose. By passing the oil into the esophageal pouch the animals are essentially avoiding systemic exposure to the fluid.

Left: *Perga affinis* larvae, aggregated, shown shortly after they were disturbed. Regurgitated oil is visible on their mouthparts and in the form of streaks on their backs. Right: A fully grown larva beside a replete esophageal sac, excised from a larva of the same size.

The entire esophagus, together with its diverticulum, is lined with a thin impervious membrane, an inner extension of the outer cuticle of the animal, and a barrier to the absorption of any materials contained within the esophagus. Trapped within the pouch and building up in quantity as the larvae grow, the oil is harmlessly stored away. It is, however, kept in readiness for defensive use. When a larva is attacked, it regurgitates some of the oil, which emerges as a viscous ooze from the mouth. Predators such as ants are effectively deterred, as are birds and mice. Pergine larvae have pulled a double evolutionary ploy. They have succeeded in circumventing their host's defense, while at the same time putting that defense to use.

Eucalypt-feeding sawflies show some variability in behavior. *Perga affinis,* one of the largest and best-known species, lives in aggregations as a larva. At night the members of a group disperse to feed, only to reaggregate in the daytime. When disturbed, the larvae respond collectively by flipping their front ends backward over their bodies, while at the same time raising the rear, in the process smear-

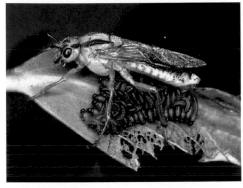

Pseudoperga guerini. Top and middle: A female guarding her eggs (laid into the blade of the leaf, and hence hidden from view), and, at a later time, her newly emerged young. Bottom: Larvae aggregated in typical fashion, with their heads directed outward, regurgitating collectively in response to disturbance.

ing themselves with oil. The term "spitfires," used in Australia to describe such larvae, is appropriate.

P. affinis larvae do not relinquish the oil until they pupate. They pupate in the soil in groups and construct contiguous cocoons. They fashion the cocoon wall out of silk and impregnate it with regurgitated oil. The oil provides the enclosed pupae with chemical protection, and as it hardens gradually over time, with physical protection as well.

Another pergine species, *Pseudoperga guerini,* differs in that it does not generally smear itself with regurgitated oil when disturbed, but simply allows the fluid to build up around the mouth. If disturbed when aggregated, *P. guerini* larvae tend to discharge collectively. Typically they arrange themselves in a circle with their mouths pointed outward, so that the entire cluster is flanked by oil. No sooner is an enemy deterred than the larvae reingest the oil. Early on, in the days immediately following emergence from the eggs, the larvae rely also on being defended by the mother, who remains in attendance during this period, ready to protect her brood against eventualities. Older larvae in some pergine species may actually imbibe more oil than needed to fill the pouch; they may void this excess by regurgitating it at times when they are not feeding.

Other sawflies have evolved very much the same strategy as the Perginae but feed on conifers instead of eucalypts. In lieu of an oil, these larvae, members of the family Diprionidae, have to deal with conifer resin, which they too separate from ingested leaf tissue and retain in the esophagus. But these larvae have a pair of esophageal pouches for accommodation of their defensive material, instead of the single pouch possessed by Perginae. Some diprionid sawflies, such as *Neodiprion sertifer,* are gregarious as larvae. They pupate individually, however, and do not incorporate the resin into their silken cocoon. Instead, during the weeks that they remain in a larva-like condition inside the cocoon before finally pupating, they use the resin to counter the efforts of any enemies that might attempt to breach the cocoon from the outside.

The regurgitant of conifer-feeding sawflies has been shown to be essentially conifer resin. In *N. sertifer,* for example, which feeds on Scotch pine *(Pinus sylvestris),* the regurgitant contains the same

A larva of *Neodiprion sertifer* regurgitating.

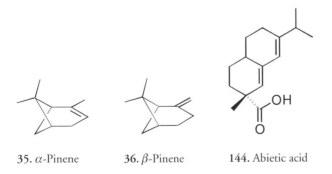

35. α-Pinene   36. β-Pinene   144. Abietic acid

blend of aromatic isoprenoids, for instance α-pinene (35) and β-pinene (36), and resin acids, for instance abietic acid (144), as the host's resin.

Plant resins are used for defensive purposes by other insects as well (see Chapter 23). Certain Australian moths of the genus *Myrascia,* for example, protect themselves as larvae by regurgitating an oil that they sequester from their foodplants (species of Myrtaceae) and store within an esophageal pouch. The parallel with pergine sawflies is striking.

Sawflies belong to the hymenopteran suborder Symphyta, which includes some 4,700 species in about 350 genera. Within the suborder, the families Pergidae and Diprionidae contain, respectively, about 300 and 50 described species.

REFERENCES

Boland, D. J., I. J. Brophy, and A. P. N. House. 1997. *Eucalyptus Leaf Oils: Use, Chemistry, and Marketing.* Melbourne: Inkata Press.

Carne, P. B. 1962. The characteristics and behavior of the sawfly *Perga affinis affinis* (Hymenoptera). *Australian Journal of Zoology* 10:1–34.

———— 1969. On the population dynamics of the eucalypt-defoliating sawfly *Perga affinis affinis* Kirby (Hymenoptera). *Australian Journal of Zoology* 17:113–141.

Common, I. F. B., and T. E. Bellas 1977. Regurgitation of host-plant oil from a foregut diverticulum in the larvae of *Myrascia megalocentra* and *M. bracteatella* (Lepidoptera; Oecophoridae). *Journal of the Australian Entomological Society* 16:141–147.

Eisner, T., J. S. Johnessee, J. Carrel, L. B. Hendry, and J. Meinwald. 1974. Defensive use by an insect of a plant resin. *Science* 184:996–999.

Langenheim, J. H. 2003. *Plant Resins: Chemistry, Evolution, Ecology, and Ethnobotany.* Portland, Ore.: Timber Press.

Morrow, P. A., T. E. Bellas, and T. Eisner. 1976. *Eucalyptus* oils in the defensive oral discharge of Australian sawfly larvae (Hymenoptera: Pergidae). *Oecologia* 24:193–206.

Penfold, A. R., and J. L. Willis. 1961. *The Eucalypts.* London: Academic Press.

Schmidt, S., G. H. Walter, and C. J. Moore. 2000. Host plant adaptations in myrtaceous-feeding pergid sawflies: essential oils and the morphology and behaviour of *Pergagrapta* larvae (Hymenoptera, Symphyta, Pergidae). *Biological Journal of the Linnean Society* 70:15–26.

Welch, M. B. 1920. *Eucalyptus* oil glands. *Journal of the Proceedings of the Royal Society of New South Wales* 54:208–217.

68

Class INSECTA

Order HYMENOPTERA

Family Formicidae

Camponotus floridanus

A carpenter ant

A *Camponotus floridanus* worker beside
the formic acid–bearing sac excised from
another worker of comparable size.

Ants are numerous and occur virtually everywhere. Nearly 9,000
species are known, but many doubtless remain to be discovered.
Their abundance and diversity defy the imagination. For example, a
single hectare of savanna in the Ivory Coast has been estimated to
contain some 20 million individual ants, while as many as 43 separate species have been collected from a single tree in the Peruvian
Amazon. Ants are of immense ecological importance, as consumers
of insects, tillers and modifiers of the soil, and promoters of plant

dispersal. Rightly, they have been called the paragons of the insect world.

Ants are primarily soil dwellers, living in perennial colonies, in which the queen is the reproductive entity, and the workers the sterile, flightless daughter caste, which performs most colony growth and maintenance activities, including societal defense. The worker caste is sometimes divided into physically distinct subcastes, named according to size major workers and minor workers. Males are short-lived, and are produced in scant numbers in the mature colony only, when needed for reproductive purposes.

To bear witness to the efficacy of colony defense in ants, all you need to do, in many species, is to disturb the workers that mount guard at the entrance of a nest. Poke these individuals or tap the ground beside them, and you are likely to find yourself stung, bitten, smeared with secretion, or even sprayed by the ants, which converge upon the site of trouble as if rallied by an assembly call. Not all ants react as dramatically, although nest defense by coordinated worker action is commonplace in ants. The defense, of course, also serves the individual worker ant, when it is challenged while on solitary assignment at a distance from the nest.

Many ants depend on venoms for defense. Like social bees and wasps, they administer the poison by injection, using a stinging apparatus derived evolutionarily from the ovipositor (egg-laying organ) of their progenitors. Unlike social bees and wasps, however, which use their sting almost exclusively for defense, ants use the sting both for protection and for incapacitation of prey.

Stinging prevails in the anatomically more primitive ants, including most Ponerinae and Myrmicinae, as well as in the Myrmeciinae, Nothomyrmeciinae, Aneuretinae, Pseudomyrmecinae, Dorylinae, and Ecitoninae.

Ant venoms are complex mixtures, unusual in that they may contain high concentrations of alkaloids. Thus, for instance, in *Solenopsis invicta,* the notorious fire ant, a member of the ant subfamily Myrmicinae, the venom is composed primarily of piperidine alkaloids (2-alkyl or alkenyl 6-methyl piperidines) (for example, **145**), toxins that act by triggering the release of histamine and other vasoactive amines from mast cells. Such action can account for the

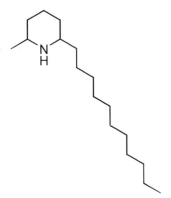

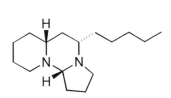

145. 2-Methyl-6-undecylpiperidine

146. Tetraponerine-8

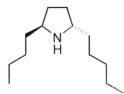

147. *(2S,5S)-2-Butyl-5-pentylpyrrolidine*

sharp localized pain you feel if stung by this ant. Fire ant alkaloids are also said to inhibit the synthesis of nitric oxide, a neurotransmitter centrally involved in a number of physiological and inflammatory processes in vertebrates. When stinging, fire ants attempt simultaneously to bite. By clamping down with the mandibles they secure the hold they need to jab the sting into the enemy.

Fire ants also resort to aerial dispersal of their venom: they raise and vibrate the abdomen, while at the same time extruding a droplet of venom from the stinger. They do this when fending off ants of other species at a feeding site, or when tending the brood. The venom has antibiotic potency, and its application in aerosol form to the brood is believed to protect against microbial infection. In the queen of *S. invicta,* the venom has pheromonal potency. It elicits orientation and attraction of workers, and affects worker attentiveness to the brood.

Isolation of venom components is currently proceeding at an accelerated pace and has resulted in the characterization of interesting alkaloids. Among these compounds are, for instance, the tetraponerines (for example, **146**) from species of *Tetraponera* (subfamily Pseudomyrmecinae) from New Guinea. These compounds are emitted by the ants as part of a venomous paste, which the ants smear onto the surface of the enemy by use of their modified sting. Among alkaloids administered by stinging are certain 2,5-dialkylpyrrolidines (for example, **147**) produced by *Monomorium indicum* (subfamily Myrmicinae).

A familiar North American species that has been studied in some detail is *Pogonomyrmex badius* (subfamily Myrmicinae), a harvester ant, which produces a venom rich in enzymes, mostly phospholipases and hyaluronidase. Such pain-inducing compounds occur also in other insect venoms, though at lower concentrations. Also produced by this ant is a lipase and small quantities of histamine. The venom is highly toxic to vertebrates (and extremely painful to humans) and appears fashioned to serve primarily for defense against vertebrates. In insects the venom appears to act primarily by virtue of its histolytic properties.

Remarkably, horned lizards of the genus *Phrynosoma* can feed on *Pogonomyrmex* ants. They sometimes sit beside the entrance of a *Pogonomyrmex* nest for protracted periods, flipping the ants into the mouth with the tongue, one after the other, as the ants walk by on their way to and from the nest. *Phrynosoma* possess a factor in the blood plasma that makes them resistant to *Pogonomyrmex* venom.

Also of interest is that frogs appear to tolerate the toxic components of ant venoms, as they do the active principles of the defensive secretions of many other arthropods. These chemicals, after ingestion by the frogs, instead of being quickly eliminated with the feces or urine, appear in at least some cases to settle in the frogs' skin, where they may act to convey protection upon the frog itself. Among the substances thus used for defensive purposes by frogs are alkaloids known to be, or presumed to be, derived from ant venoms. Both piperidines (**145**) and pyrrolidines (**147**), for example, have been detected in frog skin. So have certain pumiliotoxins (for example, **148**), which had not previously been known from arthropods,

Left: A horned lizard *(Phrynosoma cornutum)*. Note the ant head *(Pogonomyrmex* species) clamped to the animal's neck. Right: *Pogonomyrmex* ant remains from a fecal sample of a *Phrynosoma* lizard.

148. Pumiliotoxin-8

but have now been isolated from ants that co-occur with the frogs. The frogs in question include the "poison dart frogs" of the family Dendrobatidae, the skin secretions of which are used to this day by indigenous tribes in the American tropics to poison the tips of the spears they use in hunting small birds and game.

Ants of one of the most successful subfamilies, the Formicinae, have taken the modification of the ovipositor one step further. They still possess a poison gland, but they have no stinger associated with it. Instead of injecting their secretion, they deliver the fluid by forcible ejection, their gland having been transformed to function essentially as a spray gun. In Formicinae the glands have also shifted from the production of venoms to the production of one major compound, an irritant, meant to be administered topically. That com-

H—C(=O)—OH

68. Formic acid

pound, formic acid (**68**), the simplest of all carboxylic acids, was one of the first natural products ever characterized from arthropods, having been isolated in pure form from the distillate of formicine ants as early as 1670. Formicid spray may contain upward of 50% formic acid (see Chapter 34). The sac in which the fluid is stored is sizeable, and when replete can take up much of the volume of the worker ant's abdomen. Accommodation of a large acid sac poses no problem for workers, because they usually lack a fully developed reproductive system. As has been shown by injection of radio-labeled precursors, formicine ants synthesize formic acid themselves, from the amino acids L-serine and glycine.

Formic acid, at the high concentrations in which it occurs in the storage sac, is potently repellent. But the compound is water soluble, a drawback that prevents it from penetrating readily through the lipid-coated skeletal cuticle of insects. Formicine ants apparently solve this problem by biting at the same time they spray. By biting they perforate or abrade the cuticle of target insects, thereby creating routes for penetration of the acid. In fact, when they spray, they usually bend the abdomen forward beneath the body, to direct the fluid onto the site being bitten. The bites may themselves be effectively defensive, against both arthropods and vertebrates. Naturalists are quick to learn that if they turn over a log and uncover a nest of carpenter ants (species of the genus *Camponotus*), they should not reach directly into that nest, lest they be prepared to bear first the pain induced by the workers' bites and then the burning sensation elicited when the acid spray works itself into the lacerations inflicted by the bites. *Camponotus floridanus,* an ant commonly found throughout Florida, provides a good example of a species fiercely defensive of its nest. An ideal experimental animal, *C. floridanus* is easy to maintain in the laboratory, where it is as doggedly protective of its colonies as it is in the field.

326

Left: A *Camponotus floridanus* worker biting while bending the abdomen forward to direct the spray toward the site bitten. Right: Another *C. floridanus* worker biting into a piece of rubber and spraying onto indicator paper, where the acid spray is registered in white. The ant withheld its spray until it was given the opportunity to sink the mandibles into the piece of rubber.

It is common for ants, when threatened, to alert nestmates to the state of emergency by emitting alarm pheromones. Alarm pheromones are produced by stinging and spraying ants alike. In ants that spray formic acid, the spray itself may provide the alarm signal in some species. But more commonly, in ants generally, alarm pheromones are produced in separate glands, often as part of multifunctional blends of compounds. These glands, which appear to have arisen independently several times in the course of ant evolution, have been variously named (for instance, mandibular, pygidial, cepahlic, and Dufour's glands), in recognition of either their anatomical location or their discoverer. Alarm pheromones are fundamental to the coordination of group defense in ants, just as they are for comparable purposes in other social Hymenoptera and termites (see Chapters 18 and 69).

One ant in which the integration of defensive and alarm behavior has been studied is *Acanthomyops claviger*, a member of the Formicinae. Like formicines generally, this ant makes use of formic acid, which it produces in its poison gland and discharges with its spray. In addition, it depends on its mandibular glands, two large sacs opening at the base of the mandibles and producing a mixture of short-chain unsaturated aldehydes (including the isoprenoid, citronellal; **149**), and its Dufour's gland, a sac opening beside the poison gland on the abdominal tip, and secreting a mixture of

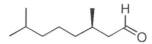

149. D-Citronellal

An ant-mimicking spider, *Synemosyna formica.*

straight-chain saturated hydrocarbons and ketones. The substances produced by these glands all have defensive capacity vis-à-vis insects, and they are discharged when the ant is in conflict with other insects. Formic acid and citronellal alone, are potently repellent and could in themselves provide for a significant level of protection. But the compounds from the mandibular and Dufour's glands have strong alarming potential as well, and can therefore act in an important way to bring workers to the aid of distressed nestmates. Indeed, the *A. claviger* worker discharges from these glands promptly when disturbed, meeting the challenge head on and calling for help before the "problem" gets out of hand. It has been estimated that the pheromonal call of an *A. claviger* worker can summon nestmates from as far as 10 centimeters away.

Quite remarkable is the defense of a tropical Asian formicine ant (of the *Camponotus saundersi* group) in which the workers offer protection by literally blowing themselves up. The mandibular glands of this ant are hypertrophied, so that they take up not only much of the head, but also much of the abdomen. When the ants are assaulted, they compress the abdomen abruptly to the point that it bursts, casting the sticky contents of the mandibular glands in all directions. The trade-off in this altruistic suicide is that the attackers become trapped by the glue.

Given the effectiveness of their defenses, ants have relatively few enemies. Not surprisingly, there are arthropods that mimic ants and are therefore avoided by predators that shun ants. Among such ant mimics is the jumping spider, *Synemosyna formica*.

REFERENCES

Braekman, J. C., D. Daloze, J. M. Pasteels, P. van Hecke, J. P. Declercq, V. Sinnwell, and W. Francke. 1987. Tetraponerine 8, an alkaloidal contact poison in a Neo-Guinean pseudomyrmecine ant *Tetraponera* sp. *Zeitschrift fuer Naturforschung* 42C:627–630.

Daly, J. W. 1995. The chemistry of poisons in amphibian skin. *Proceedings of the National Academy of Sciences USA* 92:9–13.

——— 1998. Thirty years of discovering arthropod alkaloids in amphibian skin. *Journal of Natural Products* 61:162–172.

Daly, J. W., H. M. Garraffo, and T. F. Spande. 1999. Amphibian alkaloids: chemistry, pharmacology, and biology. In S. W. Pelletier, ed., *Alkaloids: Chemical and Biological Perspectives*. Oxford: Pergamon.

Daly, J. W., H. M. Garraffo, T. F. Spande, V. C. Clark, J. Ma, H. Ziffer, and J. F. Cover, Jr. 2003. Evidence for an enantioselective pumiliotoxin 7-hydroxylase in dendrobatid poison frogs of the genus *Dendrobates*. *Proceedings of the National Academy of Sciences USA* 100:11092–11097.

Hölldobler, B., and E. O. Wilson. 1990. *The Ants*. Cambridge, Mass.: Belknap Press of Harvard University Press.

——— 1994. *Journey to the Ants*. Cambridge, Mass.: Belknap Press of Harvard University Press.

Jones, T. H., M. S. Blum, P. Escoubas, and T. M. M. Ali. 1989. Novel pyrrolidines in the venom of the ant *Monomorium indicum*. *Journal of Natural Products* 52:779–784.

Jones, T. H., M. S. Blum, A. N. Andersen, H. M. Fales, and P. Escoubas. 1988. Novel 2 ethyl-5-alkylpyrrolidines in the venom of an Australian ant of the genus *Monomorium*. *Journal of Chemical Ecology* 14:35–46.

Leclerq, S., J. C. Brackman, J. C. Daloze, and J. M. Pasteels. 2000. The
defensive chemistry of ants. *Progress in the Chemistry of Organic Natural Products* 79:115–129.

Lind, N. K. 1982. Mechanism of action of fire ant *(Solenopsis)* venoms. I.
Lytic release of histamine from mast cells. *Toxicon* 20:831–840.

Merlin, P., J. C. Braekman, D. Daloze, and J. M. Pasteels. 1988.
Tetraponerines, toxic alkaloids in the venom of the Neo-Guinean
pseudomyrmecine ant *Tetraponera* sp. *Journal of Chemical Ecology*
14:517–528.

Obin, M. S., and R. K. Vander Meer. 1985. Gaster flagging by fire ants
(*Solenopsis* spp.): functional significance of venom dispersal behavior.
Journal of Chemical Ecology 11:1757–1768.

Read, G. W., N. K. Lind, and C. S. Oda. 1978. Histamine release by fire
ant *Solenopsis* venom. *Toxicon* 16:361–368.

Schmidt, J. O., and M. S. Blum. 1978. The biochemical constituents of
the venom of the harvester ant *Pogonomyrmex badius. Comparative
Biochemistry and Physiology* 61C:239–248.

——— 1978. A harvester ant venom chemistry and pharmacology. *Science* 200:1064–1066.

——— 1978. Pharmacological and toxicological properties of harvester
ant *Pogonomyrmex badius* venom. *Toxicon* 16:645–652.

Schmidt, J. O., M. S. Blum, and W. L. Overal. 1986. Comparative
enzymology of venoms from stinging Hymenoptera. *Toxicon*
24:907–922.

Schmidt, P. J., W. C. Sherbrooke, and J. O. Schmidt. 1989. The detoxification of ant *Pogonomyrmex* venom by a blood factor in horned lizards *Phrynosoma. Copeia* 1989:603–607.

Vander Meer, R. K., B. M. Glancey, C. S. Lofgren, A. Glover, J. H.
Tumlinson, and J. Rocca. 1980. The poison sac of red imported fire
ant *Solenopsis invicta* queens source of a pheromone attractant. *Annals of the Entomological Society of America* 73:609–612.

Wilson, E. O. 1971. *The Insect Societies.* Cambridge, Mass.: Harvard University Press.

Yi, G. B., D. McClendon, D. Desaiah, J. Goddard, A. Lister, J. Moffitt,
R. K. Vander Meer, R. K. deShazo, K. S. Lee, and R. W. Rockhold.
2003. Alkaloid component of *Solenopsis invicta* (fire ant) venom inhibits nitric oxide synthesis. *International Journal of Toxicology*
22:81–86.

69

Class INSECTA

Order HYMENOPTERA

Family Apidae

Apis mellifera

The honey bee

A honey bee worker, in alerted stance, with the stinger extruded.

Of all the insects that inspire fright, none are more dreaded than the stinging Hymenoptera, especially the bees and wasps. Although some Hymenoptera are in fact stingless, such as the sawflies (see Chapter 67) and certain bees and ants, that does not prevent humans from extending their dislike to the Hymenoptera as a whole. The threat posed by these insects is real. Bee stings and wasp stings are painful, and in humans allergic to the venom they can induce life-threatening reactions. It is ironic that the most dangerous stinging insect should be the honey bee, a species that, both as pollinator and honey producer, has been of such extraordinary service to humans. More individuals in the United States die as a consequence of honey bee stings than from the bites or stings of any other venomous animals, including spiders, scorpions, and snakes.

The stinging apparatus of the Hymenoptera is restricted to females, a reflection of the fact that the device is evolutionarily derived from the ovipositor (egg-laying organ). In the honey bee society, the stinging apparatus is a standard feature of the workers, the sterile females that make up the bulk of the society. The worker bees perform all basic activities in the hive, from nectar and pollen collection to brood care and defense. The stinger is essential, not only for the protection of the individual worker bee when it is on lone duty away from the hive, but for defense of the hive itself, for which purpose workers may rally in groups to consolidate their stinging action. It is best not to wander into the territory immediately around a hive, lest you provoke a massive response from the workers.

The stinging apparatus of the honey bee worker is remarkably sophisticated in design. Venom is produced by paired glands within the abdominal cavity and is stored in an expanded poison sac. The sac empties via a narrow canal into the hollow core of an elongate, apically barbed sting. When on the attack, the honey bee first jabs the sting into the victim's flesh, by means of a ballistic movement of the abdomen. But venom delivery is triggered only when the bee pulls away and leaves the stinging apparatus behind. Although the bee itself is doomed by the loss of the rear of its abdomen, what remains in the victim serves as an autonomous venom pump. The sting itself is made up of a sharp central stylet and two barbed lancets, pressed together lengthwise to create a canal for the delivery of venom. And when the pump is engaged, these parts move in coordination to drive the sting further into the flesh. The stylet and lancets are the outward extensions of robust cuticular plates within the base of the abdomen. The plates are shaped to serve as a mechanical pump that includes elaborate valves for the one-way delivery of venom, and they also have broad surfaces for the attachment of

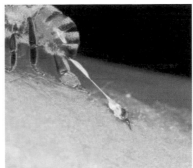

Stinging in the honey bee is initiated when the insect thrusts its stinger into the enemy (top left). The bee then pulls away, tearing the stinger apparatus from its body (top right). Left behind, embedded in the flesh of the victim, the apparatus starts the pumping action that ensures delivery of the venom (bottom).

the skeletal muscles that operate the pump. The muscles are actively driven by motor neurons contained within the most posterior part of the central nervous system, but their activity is normally prevented by inhibitory signals from more anteriorly located neurons. Once the stinging apparatus is torn from the bee inhibition is precluded, so that the musculature is free to undertake the rhythmic pumping action that drives venom into the victim.

The venom is chemically complex, and its individual components have specialized roles. Melittin, a peptide, constitutes up to half of the dry weight of the venom, and is largely responsible for the protracted pain associated with a honey bee sting. Melittin acts as an ionophore: it penetrates cell membranes, forming channels through which ions can diffuse. Pain receptor neurons, when thus affected, are activated. A second component, the enzyme hyaluronidase, acts by hydrolyzing hyaluronic acid, an important component of the intercellular binding material in connective tissue. Through its action, intercellular spaces are created through which the venom can more effectively spread. Another enzyme, a phospholipase of the A_2 type, hydrolyzes phospholipids, thereby promoting additional membrane breakdown, and potentiating the action of the hyaluronidase. The breakdown of phospholipids is of further importance in that it triggers events leading to the formation of a variety of inflammatory compounds. The phospholipase is also the major allergen in the venom, and is thus responsible for the life-threatening events that evolve in sensitive humans. The large weal produced at the site of a bee sting is characteristic of a histamine response. Histamine is present in the venom in small amounts, and may contribute to the initial sensation of pain. Larger amounts of histamine are released from mast cells in the skin of the victim, probably through the combined actions of melittin and phospholipase. Apart from melittin, other peptides are present in the venom in small amounts, including a mast cell degranulating peptide. Although the honey bee peptide does not appear to cause the release of histamine from mast cells in the victim's skin, closely related peptides in the venoms of other Hymenoptera are known to have that effect. The venom of queen bees differs from that of worker bees in that it lacks some of the components that are effective against vertebrates, but has instead higher concentrations of anti-insectan components. Queen venom seems tailored specifically to enable queens to kill competing queens.

The sting apparatus doubtless helps the honey bee in many situations. Most vertebrate predators avoid honey bees, either because they have learned to do so or because of genetic predisposition. Bears are undoubtedly exceptional in putting up with the hundreds of stings they must receive when raiding hives for their stores of honey.

If you are unlucky enough to provoke one honey bee to sting, you run the chance of being stung again and again, particularly if you

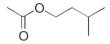

88. Isoamyl acetate

arc near the hive. When the worker bees guarding a hive sense an impending threat, they give a pheromonal alarm call. The pheromone is a volatile mixture of more than 20 compounds, the main one being isoamyl acetate (**88**). Agitated guards stand fast outside the hive, evert the sting, and rapidly fan the wings. Everting the sting exposes glandular openings at the base of the sting through which the pheromone is released, and the agitated wing fanning ensures the pheromone's rapid dispersal. As a result of the alarm call, additional worker bees are recruited from inside the hive, and they, and any other workers in the vicinity, assume a heightened state of readiness to attack intruders. The glands responsible for producing the alarm pheromone are detached along with part of the prey-embedded sting apparatus. A sting victim thus becomes not only an intruder but an additional source of a pheromonal signal inviting attack.

The fact that stinging is usually lethal to a bee has implications. The massive responses that occur when the hive is threatened can be costly, since they can lead to the sacrifice of large numbers of workers. Interestingly, there is a situation involving group defense in a bee where the workers refrain from using the sting. The bee in question, the Japanese honey bee *Apis cerana japonica,* has as one of its enemies a large hornet, *Vespa mandarinia japonica,* which preys on worker bees by capturing them at the hive entrance. The hornet is not always successful. Sometimes it is itself grasped by one of the bees and then promptly engulfed by hundreds of others. The living, buzzing ball thus formed heats up to 47°C on the inside, a temperature that the bees can withstand but the hornets cannot. None of the bees resorts to stinging, and none is lost in the encounter. No other social insects are known that resort to microwaving as a form of defense.

The venoms of wasps and bumble bees also are powerful pain inducers, but differ from honey bee venom in composition. In their case the mast cell degranulating peptides are the principal pain-inducing agents, acting via the local release of histamine. Wasp venoms also contain kinins, peptides that contribute to inflammation. Histamine is also present in the venoms, but only in lesser amounts. Hyaluronidase and phospholipases are of subsidiary importance in the actions of such venom.

The honey bee belongs to the hymenopteran suborder Apocrita, a group of some 125,000 described species, including the wasps (parasitoid, solitary, and social), bees, and ants.

REFERENCES

Banks, B. E. C., and R. A. Shipolini. 1986. Chemistry and pharmacology of honey bee venom. In T. Piek, ed., *Venoms of the Hymenoptera: Biochemical, Pharmacological, and Behavioral Aspects.* London: Academic Press.

Bechinger, B. 1997. Structure and functions of channel-forming peptides: magainins, cercropins, melittin, and alamethicin. *The Journal of Membrane Biology* 156:97–211.

Breed, M. D., E. Guzman-Novoa, and G. J. Hunt. 2004. Defensive behavior of honey bees: organization, genetics, and comparisons with other bees. *Annual Review of Entomology* 49:271–98.

Free, J. B. 1987. *Pheromones of Social Bees.* Ithaca: Cornell University Press.

Ogawa H., Z. Kawakami, and T. Yamaguchi. 1995. Motor pattern of the stinging response in the honey bee *Apis mellifera. Journal of Experimental Biology* 198:39–47.

Ono, M., I. Okada, and M. Sasaki. 1987. Heat production by balling in the Japanese honeybee, *Apis cerana japonica,* as a defensive behavior against the hornet *(Vespa mandarinia japonica). Experientia* 43:1031–1032.

Ono, M., T. Igarashi, E. Ohno, and M. Sasaki. 1995. Unusual thermal defense by a honey bee against mass attack by hornets *(Vespa mandarinia japonica). Nature* 377:334–336.

Piek, T. 1986. Historical introduction. In T. Piek, ed., *Venoms of the Hymenoptera: Biochemical, Pharmacological, and Behavioral Aspects.* London: Academic Press.

Schmidt, J. O. 1982. Biochemistry of insect venoms. *Annual Review of Entomology* 27:339–368.

Six, D. A., and E. A. Dennis. 2000. The expanding superfamily of phospholipase A_2 enzymes: classification and characterization. *Biochimica et Biophysica Acta* 1488:1–19.

Snodgrass, R. E. 1956. *Anatomy of the Honey Bee.* Ithaca: Comstock Publishing Associates.

Visscher P. K., R. S. Vetter, and S. Camazine. 1996. Removing bee stings. *Lancet* 348:301–302.

Wager, B. R., and M. D. Breed. 2000. Does honey bee sting alarm pheromone give orientation information to defensive bees? *Annals of the Entomological Society of America* 93:1329–1332.

Epilogue

What lies in store for the student of arthropod defenses? What remains to be discovered and by whom? Naturalists, most certainly, are bound to remain involved in the exploration since so many arthropods remain unknown and since with the discovery of each new species there is the possibility of uncovering a new mode of defense. Natural products chemists, too, will remain committed to the endeavor, enticed by the realistic prospect of discovering new chemicals.

But aside from broadening the list of examples, and of molecules, the study of arthropod defenses could lead to an expansion of conceptual knowledge as well. Defenses, like other features of organisms, have evolutionary histories. How exactly did arthropod defenses come into being? What subcellular preadaptations enabled arthropods to evolve the capacity to produce defensive toxicants, let alone at the high concentrations in which these compounds are sometimes stored? What makes it possible, for example, for the whipscorpion to produce 84% acetic acid without poisoning itself? Or for a daddylonglegs to synthesize and store virtually pure 1,4-benzoquinones in its glands?

Also of interest will be the continued study of how arthropods come to possess their defensive products, whether by endogenous synthesis or by appropriation from a dietary or other exogenous source. Dietary dependencies in arthropods are varied and subtle and often related to the defensive requirements of the animals. An understanding of what exactly arthropods must eat to keep their defenses operational, and how they go about satisfying that need, has the potential to help explain many of the subtle aspects of the inter-actions of these organisms with other life forms in their environ-ment.

Moreover, interesting questions about the effectiveness of arthro-pod defenses remain unanswered. How specific are the various de-fenses in their action? Which enemies are thwarted by a given de-fense, and which are unaffected? And do those that can cope with a defense make it a habit to feed on the arthropod that manifests that defense? Are defenses, and in particular chemical defenses, induc-ible? In other words, do arthropods increase the rate of production of their defensive toxicants when they are more frequently attacked? Do they economize on production of the compounds when they are in relatively safe surroundings? And what, in precise terms, is the cost of defense to an arthropod? How much metabolic energy is in-vested in the production of, say, 1,4-benzoquinones, or a carboxylic acid such as acetic acid, or an isoprenoid such as anisomorphal? At what cost does the animal produce a gland-load of such compounds, and how does this weigh against the benefits derived? How many ants can the arthropod repel per gland-load? Or, for that matter, how many birds? And what is the mode of action of defensive sub-stances? How, in molecular terms, do they effect their deterrency?

The study of arthropod defenses will remain inextricably linked to another area of study, that of insect-plant interactions, now un-dergoing vast expansion. Plants and insects share a basic problem: they are plagued primarily by insects. It should come as no surprise therefore that the two groups have independently evolved simi-lar defenses. Their chemical defenses, in particular, are illustrative of this parallel. Many of the substances produced as anti-insectan agents by insects are produced by plants as well. This opened the option for insect opportunists to appropriate defensive substances in ready-made form from plants, a cost-effective strategy that many have adopted.

Questions raised in connection with arthropod defenses can be posed with respect to plant defenses as well. Inquiries pertaining, for example, to the inducibility of defensive compounds, or to their ef-fectiveness, or biosynthesis, are as applicable to plants as they are to arthropods. So are questions relating to secondary actions of defensive chemicals. Defensive substances, both in arthropods and plants, can "backfire," in the sense that they can help enemy organisms cue in on the producers of the compounds.

Much of interest could also be learned from instances where defensive toxins from arthropods are put to use by predators. Quite extraordinary, for example, is the discovery that certain alkaloids and other toxins present in the skin of tropical frogs are derived from ingested arthropods. There is no telling how many other organisms, including perhaps even birds, benefit defensively from incorporation of ingested arthropod toxins.

Most tantalizing are the prospects of linking the study of arthropod defenses to contemporary developments in genomics. Chemosensory signaling, which is at the root of many of the defensive interactions of arthropods, appears to be subject to similar genetic underpinnings throughout the biotic world. Arthropod defense could thus provide an ideal context for the study of gene expression, as related to chemical signal reception, transduction, and amplification. The applied benefits from such study to fields as diverse as agriculture, pest management, and medicine would be considerable.

Finally, the study of arthropod defenses, inasmuch as most arthropod defenses are chemical, falls within the domain of chemical ecology, a discipline now emergent on many fronts. Chemical ecology has done much to create awareness of the chemical value of nature. Compounds in vast numbers remain to be isolated from organisms, compounds that collectively, together with their encoding genes, amount to a treasury of immense value. Arthropods alone are bound to be the repository of much of that treasury, and should for that reason, if not solely for their beauty, be worth saving. To work in chemical ecology, even within the relatively narrow domain of arthropod defenses, obligates the practitioner to speak out on behalf of nature and its preservation. To study nature without speaking in its defense is unconscionable.

How to Study Insects
and Their Kin

Discovery in the world of insects and their kin is still very much a matter of field exploration. About 1 million species of insects have so far been described, yet estimates tell us that this is far less than half the number that exists. Add to this the number of spiders, scorpions, centipedes, and millipedes that may remain unknown, animals that together with insects make up the arthropods, or "bugs," of the world, and you will come to realize that these animals, taken collectively, are indeed still a vast unknown. Think of what this means in terms of biological wonders lying in wait, in terms of new bugs and bug adaptations awaiting discovery and elucidation. Defensive strategies are bound to be brought to light, novel and diverse, and unlike any we could have imagined. Great rewards lie in store, therefore, for those who set their sights on bugs. Naturalists of the most diverse bent can join in the exploration, as can the most inexperienced urbanophile. Biophilia, after all, is in most of us, even children, and can certainly find expression in curiosity about bugs.

Here then, should you wish to take part in the exploration, are the basic tools of the trade. The total package may be priced out of reach for some, but you need not acquire the items all at once. A canvas satchel and hand lens are quite enough to get you started, plus a pair of forceps, a soupspoon, and an assortment of vials for the temporary confinement of bugs and anything else of interest. Here's a list of what you will need.

The Canvas Satchel and Its Contents

You can buy a satchel or you can make your own, fashioning it out of sailcloth, denim, or similar material. It should have an adjustable shoulder strap, plus a strap that fits around one leg to keep it from swinging about. You can partition the interior of the satchel so that you can neatly segregate its contents.

Vials

It is fine to use plastic vials like those used by drugstores for packaging pills. Such vials are likely to be available in most households as throw-aways, but if not, you might be able to buy them from your local pharmacist. Vials should not be perforated. There is enough air in them to sustain bug life for several days. It is much more important for bug survival that the vials not dry up. You should there-

Field equipment: satchel, headlamp with battery pack, and collapsible net. At the bottom right the net is shown folded, small enough to fit in the back pocket of your pants. Note that the satchel is outfitted with a shoulder strap and a shorter strap that fits around one leg.

fore always add wet cotton wads (or wet soil, when appropriate) to vials if they are to hold bugs. Keeping the vials unperforated seals in the moisture.

You may wish on occasion to preserve bugs for later dissection or detailed examination with the stereomicroscope, for which purpose you may find it convenient to preserve some specimens in alcohol. To do that, you will need to obtain some glass vials, such as are available from biological supply firms. Both ethanol and isopropyl alcohol can serve as the preserving fluids, the latter being marketed commercially as rubbing alcohol and therefore available over the counter without a prescription.

Plastic Bags

You never know what you might wish to pick up in the field and take with you for later study. Plastic bags come in handy in many ways, as for collecting soil samples or fragments of wood, beetle larvae, caterpillars, and other herbivores you may wish to take in numbers together with clippings of their foodplants.

Forceps

Ideally, you should have three kinds of forceps. One type, often referred to as soft forceps, is just that, blunt-tipped and fashioned of spring-steel, and therefore ideally suited for seizing insects without inflicting injury. Soft forceps are obtainable from biological supply companies. Another pair, used for general seizing, both of larger insects and of inert material (as when you are digging in soil, probing decaying wood, or looking under bark), could be of the ordinary kind, obtainable from any number of sources. The third pair, suitable for specialized purposes, including (in the field) manipulation of tiny insects or (indoors) insect dissection, is of a kind called watchmaker's forceps, which come in several numerical designations, according to the sharpness of the tips. The number 3 size is ideal for general purposes, and like the other watchmaker's forceps, is available from biological supply firms. Watchmaker's forceps are expensive, however, and too delicate and dangerous to be put in the hands of an unsupervised child. It's convenient to use a holster, which can be attached to your belt, to hold forceps and scissors.

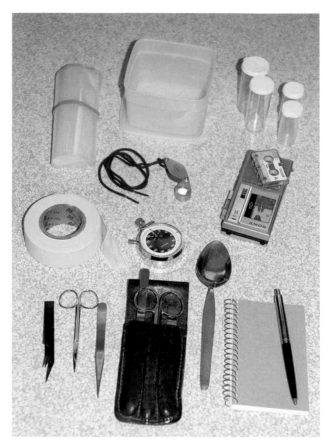

Satchel contents. Top row from the left: plastic bags; a container; and vials. Middle row: surveyor's tape; hand lens; stopwatch; dictating machine. Bottom row: holster (to be worn on a belt, for ready access to the scissors and forceps shown to its left); soupspoon; notebook and pen.

Soupspoon

This simple tool is ideal for scooping up soil, probing about in decaying wood, picking up bugs, and countless other tasks.

Stopwatch

A stopwatch is essential for timing observed behavioral events, both natural and staged. You may wish to use it in conjunction with a

Soft forceps, ideal for grasping an insect without inflicting injury.

dictating machine, to enter time-data pertinent to the descriptive narrative.

Notebook
A notebook is essential, for recording observations and data, and for making sketches.

Dictating Machine
A dictating maching is useful for keeping a descriptive record of events as they unfold.

Hand Lens (5× or 10× power)
A hand lens is essential. We make a habit of carrying it dangling from a neck strap.

Surveyor's Tape
Surveyor's tape is very convenient if you want to tag sites in the field that you might wish to revisit to make follow-up observations. To mark a site, just tie a small piece of tape to a branch or other convenient point of attachment. The tapes come in glaring colors, which makes the markers easy to locate. You should remove the tapes after

you no longer need them. They are not biodegradable and tend to be an eyesore.

Insect Net

For general use a collapsible hand net is ideal. This net, available from BioQuip Products, 17803 LaSalle Avenue Gardena, CA 90248-3602 (a firm that also markets virtually all the other items we list here) fits into your pant's back pocket when folded. It opens up by its own spring action when you pull it free and is then instantly ready for use. It takes no more than a twist of your wrist to fold it up again for pocketing.

Headlamp

Strapped to the forehead, and operated by a battery fastened to the belt, a headlamp is ideal for night exploration. Appropriate models are available in hardware and sporting goods stores.

Stereomicroscope

The stereomicroscope is the most costly item, although in many ways also the most desirable because with it you can magnify the bug to the ideal size for observation without losing three-dimensionality. More than any other device, the stereomicroscope kindles the interest of young observers, awakening in them what in the long run can develop into a passion for nature. After spending a day exploring in the field you should ideally spend that evening with the stereomicroscope, checking on the identity, behavior, and other peculiarities of the bugs you brought back in vials. You are most likely to make discoveries when the stereomicroscope reveals in detail the structural and behavioral peculiarities of the bugs you have caught.

Though expensive, stereomicroscopes are beginning to become available at more affordable prices. Good instruments are now obtainable for under $300. If you think of them as telescopes for inner space, the price will seem much more reasonable.

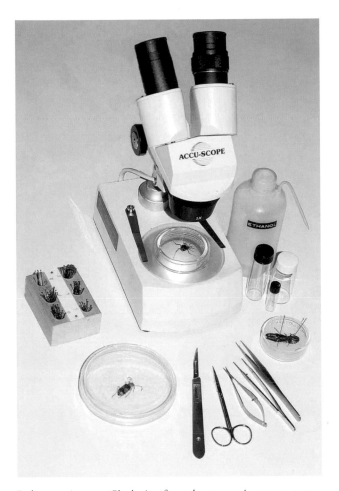

Indoor equipment. Clockwise, from the stereomicroscope: spray
bottle with ethanol; glass vials; dissecting dish (the dead insect
within is pinned to the wax bottom of the dish in readiness for
dissection); forceps, scissors, and scalpel; a petri dish bearing a
live insect; insect pins.

Depending on your level of interest, you may wish gradually to
acquire additional paraphernalia, especially petri dishes and other
containers, suitable for maintenance of insects and for observation
of their feeding behavior, courtship activities, and defensive habits.

Additional Equipment

Also useful are dissection dishes, which you can prepare by pouring melted paraffin wax into petri dishes to a depth of a few millimeters and allowing the wax to set. The wax will enable you to pin down the bug (which should be killed shortly beforehand) in the proper orientation for dissection. You should sacrifice bugs only if you have a scientific reason for doing so. You can kill bugs (usually) by placing them overnight in the freezer. Alternatively, you can kill bugs by exposing them to ethyl acetate in special killing jars (available from entomological supply companies). Dissection is best performed with the pinned bug covered with water (or better still, slightly salted water). To perform dissections you will need a fine scalpel.

Your equipment needs will vary according to your interests, and depend very much on whether you proceed beyond observation to undertake experiments, study inner anatomy, start a collection, or work with aquatic forms. Commercial catalogs may provide listings of most of the required equipment, but often what you need you can fabricate yourself. Experiments with bugs are fun, and even more so if you create the gadgetry yourself.

The Bombardier's Spray

One of the most interesting experiments you can undertake is to demonstrate the spray mechanism of bombardier beetles (Chapter 35). They are the insect world's greatest marksmen, and you can prove this nicely not only to your own satisfaction but to that of audiences of all ages.

Like other ground beetles, bombardiers have a pair of glands opening on the abdominal tip, from which they discharge an irritant secretion when disturbed. In bombardiers the secretion contains 1,4-benzoquinones and is unique in that it is discharged hot, at 100°C. To demonstrate the spray, all you need to do is tether a bombardier, place it on a special indicator paper, and subject it to simulated attack by pinching its legs with forceps, simulating an ant bite.

Bombardiers are of worldwide distribution and include a number of genera. The North American species all belong to the genus

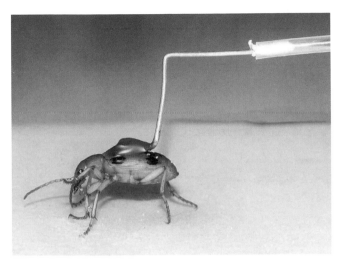

A bombardier beetle fastened to its tether.

Brachinus, and while they vary somewhat in size, ranging in length from about 5 to 15 millimeters, they are readily recognized by their blue wing covers and orange-brown body. *Brachinus* occur mostly near lakes and streams, and since they are active primarily at night, you can sometimes capture them in light traps or locate them with the help of headlamps as they scurry about on the ground.

Brachinus are long-lived and can be maintained on a diet of freshly cut up insects (for example mealworms, the larvae of the beetle *Tenebrio molitor*) and water (offered by way of soaked cotton wads). These beetles are not cannibalistic and you can therefore keep them in groups. Their most vital requirement is moisture. They cannot long withstand confinement in a hot dry enclosure. Properly maintained, bombardiers can survive for well over a year.

To prepare a bombardier beetle for testing, you must first attach a small wire hook to its back (either to the anterior part of the thorax or the base of the wing covers) with wax. The hook provides a convenient handle with which you can grasp the beetle without inducing it to discharge, and with which you can link it (using a small piece of plastic tubing) to one end of a tethering rod (a wooden applicator stick or piece of wire will do nicely). Fasten the other end of the rod to a vertical stand, preferably with a universal clamp.

The best wax with which to attach the hook to the beetle is Sticky Wax (Kerr Laboratories, Emeryville, CA, 94608), widely used in dentistry (you should have no problem talking your dentist into giving you a sample). Beeswax and paraffin waxes will also do, but they don't stick nearly as well to the beetle as Sticky Wax. To prevent the beetle from discharging when it is being outfitted with the hook, cool it down beforehand by releasing it into a dish with sand that had been stored in the freezer ($-4°C$). The beetle is immobilized within a minute after being placed on such a cold substrate, and you can then attach the hook without difficulty. The beetle will regain full mobility within a few minutes after being returned to ambient temperature, and it is then ready to "perform."

To prepare the sheet of indicator paper, soak a piece of conventional filter paper (15 × 15 centimeters) in a special solution mixed in an open pan (a conventional 21 × 21 centimeter brownie baking pan works well). The solution discolors spontaneously upon exposure to air, so you must mix it within an hour of using it. Prepare the solution by mixing water (150 milliliters) and cornstarch (10 grams) in the pan, then adding potassium iodide (6 grams) and concentrated hydrochloric acid (about 6 milliliters) while stirring. Then soak the sheet of paper in the fluid, lay it out on a glass plate, and blot off the excess liquid with a paper towel. It is then ready for use.

To demonstrate the spraying mechanism, place an individual beetle that you have already fastened to its tethering rod on the center of the paper sheet, taking care to adjust the rod so that the beetle is oriented in a normal stance. Then stimulate the beetle by pinching its legs with forceps, one at a time, simulating what might occur in an ant attack. Each pinching is likely to elicit a discharge. First you will hear a distinctly audible "pop," and then almost instantaneously, as if by magic, you will see the dark brown spray pattern appear on the paper. You can usually elicit over 10 discharges from a beetle that has gone unmolested for several days. After the experiment, you can easily detach the beetle from the hook and return it to its cage (or to nature).

You can use this technique with other bombardiers as well, including, besides the larger tropical relatives of *Brachinus* (for exam-

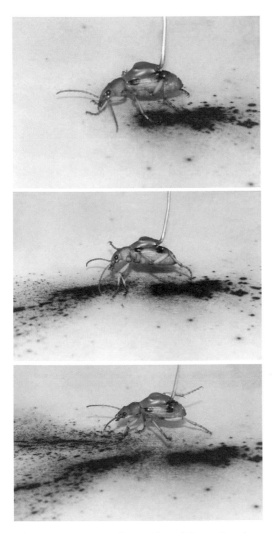

Three consecutive discharges elicited from a bombardier beetle, showing the spray patterns left on indicator paper. The beetle was stimulated by pinching, in sequence, its left hindleg, left foreleg, and right foreleg. Note that the three discharges were accurately aimed.

ple, the species of *Pheropsophus* and *Stenaptinus*), all other members of the subfamily Ozaeninae.

You can also use indicator papers of different kinds to demonstrate spray aiming in other ground beetles. Thus, for example, you can use phenolphthalein-impregnated paper to depict the formic acid–containing spray of species of *Galerita* (Chapter 34). The impregnating fluid in this case is a 1% solution of phenolphthalein in 70% aqueous ethanol, to which you add a small amount of crystalline potassium hydroxide to turn the fluid red. The acid spray of the beetle registers in white on the paper. You can also use this paper to demonstrate the acetic acid–containing spray of the whipscorpion (Chapter 1).

Acknowledgments

We wish to express our thanks to the many Cornell graduate students, undergraduates, postdoctoral associates, and technical assistants who participated over the years in diverse capacities, some as biologists, others as chemists, in our studies of animal defenses. Special thanks are due our friend and longtime collaborator Jerrold Meinwald, who as expert chemist guided the efforts to isolate, characterize, and synthesize the numerous substances that we know now are responsible for the effectiveness of these defenses.

Support for our work came from various sources. We are especially indebted to the National Institutes of Health, which has backed our Cornell studies uninterruptedly since 1959, and to Johnson & Johnson, which provided generous funding in recent years. One of us (M.S.) is grateful to Johnson & Johnson for support received during the year she spent at Cornell working on the book, while on sabbatical leave from Emory University. We are very grateful to Matthew Gronquist and Jerrold Meinwald for checking the chemical formulas for accuracy, and to Dr. Gronquist for drawing the formulas.

We are much indebted to colleagues and friends in our respective institutions, and from throughout the world, for courtesies and professional favors extended to us over the years. Of particular value has been the help received from the staff of University Photography at Cornell and from the Cornell libraries. Janis Strope, of our Cornell offices, provided invaluable aid in the preparation of the manuscript.

Some of the work discussed herein was carried out at the Archbold Biological Station, Lake Placid, Florida. We are beholden to the entire staff of that wonderful institution for having provided us with the opportunity, over a period of nearly five decades, to explore the vast biological unknown that lies hidden within its holdings.

We owe thanks to William A. Shear of Hampden-Sydney College for sharing his knowledge on opilinoids, and Désiré Daloze and Jacques M. Pasteels of the University of Brussels for information pertinent to the chemistry of ants. Helen Ghiradella of the State University of New York at Albany provided many insightful comments on the text.

Working with Harvard University Press on the production of this book has been a sheer pleasure. For countless favors rendered, we thank Michael Fisher, Editor-in-Chief, and Sara Davis, as well as Annamarie Why, who designed the book, and David Foss, who guided it through the production phases. Nancy Clemente edited the entire manuscript, proving once again that she is not only the best of critics, but unquestionably also the kindest.

Illustration Credits

Some of the illustrations, listed next below by source, have appeared previously in papers published originally in the journal *Science*. These are reproduced here with permission from the American Association for the Advancement of Science.

Page 50	Smolanoff, J., A. F. Kluge, J. Meinwald, A. McPhail, R. W. Miller, K. Hicks, and T. Eisner. 1975. *Science* 188:734–736.
Page 75	Eisner, T. 1958. *Science* 128:148–149.
Pages 121, 123, 149 (bottom)	Eisner, T., K. Hicks, M. Eisner, D. S. Robson. 1978. *Science* 199:790–794.
Page 135 (top)	Eisner, T., S. Nowicki, M. Goetz, and J. Meinwald. 1980. *Science* 208:1039–1042.
Page 187 (right)	Meinwald, J., Y. C. Meinwald, A. M. Chalmers, and T. Eisner. 1968. *Science* 160:890–892.
Page 238 (left, top right)	Eisner, T. and J. Meinwald. 1966. *Science* 153:1341–1350.

We are also grateful for permission to reproduce the following figures, which appeared initially in the *Proceedings of the National Academy of Sciences USA*:

Pages 37, 39	Eisner, T., M. Eisner, A. B. Attygalle, M. Deyrup, and J. Meinwald. 1998. *Proc. Nat. Acad. Sci. USA* 95:1108–1113.

Page 53 (top)	Carrel, J. E. and T. Eisner. 1984. *Proc. Nat. Acad. Sci. USA* 81:806–810.
Pages 56 (left), 57 (right)	Eisner, T., M. Eisner, and M. Deyrup. 1996. *Proc. Nat. Acad. Sci. USA* 93:10848–10851.
Page 145 (right)	Eisner, T., A. B. Attygalle, W. E. Conner, M. Eisner, E. MacLeod, and J. Meinwald. 1996. *Proc. Nat. Acad. Sci. USA* 93:3280–3283.
Pages 152 (left), 155	Rossini, C., A. B. Attygalle, A. Gonzalez, S. R. Smedley, M. Eisner, J. Meinwald, and T. Eisner. 1997. *Proc. Nat. Acad. Sci. USA* 94:6792–6797.
Page 166	Eisner, T. and D. J. Aneshansley. 2000. *Proc. Nat. Acad. Sci. USA* 97:11313–11318.
Page 169	Meinwald, J., Q. Huang, J. Vrkoc, K. B. Herath, Z-C. Yang, F. Schröder, A. B. Attygalle, V. K. Iyengar, R. C. Morgan, and T. Eisner. 1998. *Proc. Nat. Acad. Sci. USA* 95:2733–2737
Pages 190, 192	Eisner, T., M. A. Goetz, D. E. Hill, S. R. Smedley, and J. Meinwald. 1997. *Proc. Nat. Acad. Sci. USA* 94:9723–9728.
Page 207	Schroeder, F. C., S. R. Smedley, L. K. Gibbons, J. J. Farmer, A. B. Attygalle, T. Eisner, and J. Meinwald. 1998. *Proc. Nat. Acad. Sci. USA* 95:13387–13391.
Pages 224, 226	Eisner, T., S. R. Smedley, D. K. Young, M. Eisner, B. Roach, and J. Meinwald. 1996. *Proc. Nat. Acad. Sci. USA* 93:6494–6498.
Pages 247, 248	Eisner, T. and D. J. Aneshansley. 2000. *Proc. Nat. Acad. Sci. USA* 97:6568–6573.
Page 151 (right)	Eisner, T. and M. Eisner. 2000. *Proc. Nat. Acad. Sci. USA* 97:2632–2636.
Page 289	Eisner, T. and J. Meinwald. 1995. *Proc. Nat. Acad. Sci. USA* 92:50–55.

Page 306   Smedley, S. R., F. C. Schroeder, D. B. Weibel, J.
                  Meinwald, K. A. LaFleur, J. A. Renwick, R.
                  Rutowski, and T. Eisner. 2002. *Proc. Nat. Acad.
                  Sci. USA* 99:6822–6827.

Page 327 (left)   Eisner, T. , I. T. Baldwin, and J. Conner. 1993.
                  Proc. Nat. Acad. Sci. USA 90:6716–6720.

Thanks also go to the following publishers for permission to reproduce figures from some of their journals.

Entomological Society of America

Page 128   Eisner, T. and M. Eisner. 1989. *Bull. Entomol.
                  Soc. Amer.* 35:9–11.

Springer Verlag

Pages 137, 138   Mason, R. T., H. M. Fales, M. Eisner, and T.
(right)   Eisner. 1991. *Naturwissenschaften* 78:28–30.

Page 12 (right)   Eisner, T., D. Alsop, and J. Meinwald. 1978.
                  Arthropod Venoms, ed. S. Bettini, Handbook of
                  Experimental Pharmacology, vol. 48, pp. 87–99.
                  Berlin: Springer-Verlag. Berlin.

Pages 273, 274   Darling, D. C., F. Schroeder, J. Meinwald, M.
                  Eisner, and T. Eisner. 2001. *Naturwissenschaften*
                  88:306–309.

Pages 84, 86   Eisner, T., I. Kriston, and D. J. Aneshansley.
                  1976. *Behavioral Ecology and Sociobiology* 1:83–
                  125.

Pages 316, 317   Morrow, P. A., T. E. Bellas, and T. Eisner. 1976.
(bottom)   *Oecologia* 24:193–206.

The Company of Biologists

Page 9   Eisner, T., C. Rossini, A. González, and M.
                  Eisner. 2004. *Journal of Experimental Biology*
                  207:1313–1321.

Page 91   Eisner, T., R. C. Morgan, A. B. Attygalle, S. R.
                   Smedley, K. B. Herath, and J. Meinwald. 1997.
                   Journal of Experimental Biology 200:2493–2500.

Kluwer Academic/Kluwer Publishing

Page 143   Masters, W. M. and T. Eisner. 1990. *Journal of
                   Insect Behavior* 3:143–157.

Page 179   Jefson, M., J. Meinwald, S. Nowicki, K. Hicks,
                   and T. Eisner. 1983. *Journal of Chemical Ecology*
                   9:159–180.

Page 271   Attygalle, A. B., S. R. Smedley, J. Meinwald, and
                   T. Eisner. 1993. *Journal of Chemical Ecology*
                   19:2089–2104.

Birkhauser Verlag AD

Pages 77, 79   Eisner, T., C. Rossini, and M. Eisner. 2000.
(bottom)   *Chemoecology* 10:81–87.

The Lepidopterist's Society

Pages 261, 263   Epstein, M., S. R. Smedley, and T. Eisner. 1994.
(bottom)   *Journal of the Lepidopterist's Society* 48:381–386.

Sources of other illustrations are:

Page 159   Beetle outline courtesy of Frances Fawcett

Page 333   Kirk Visscher

Pages 15, 29   Noel Snyder

Page 89   Milan Busching

Page 160   Thomas Eisner and Daniel Aneshansley

Index

carpenter bee, 205

Catharus ustulatus (thrush), 174

centipedes. *See* Chilopoda

Centruroides exilicauda (bark or deadly sculptured scorpion), 16, 18. *See also* Scorpiones

Ceraeochrysa (green lacewings): *cubana*, 141–144; *lineaticornis*, 149; *smithi*, 128, 145–147. *See also* Neuroptera: Chrysopidae

Ceratiola ericoides (rosemary bush), 283–284

Chauliognathus (soldier beetles): *lecontei*, 185–188; *pennsylvanicus*, 186

Chelimorpha cassidea (Argus tortoise beetle), 251–253. *See also* Coleoptera: Chrysomelidae

Chelinidea vittiger (leaf-footed bug), 102–107

Chilopoda (centipedes): Geophilida (soil), 31, 33–36; Lithobiida (stone), 31, 33; Scolopendrida (giant), 29 32; Scutigerida (scutigerid), 31

Chrysomela scripta (cottonwood leaf beetle), 257–258. *See also* Coleoptera: Chrysomelidae

chrysomelidial, 257

Chrysopa slossonae (green lacewing), 148–150. *See also* Neuroptera: Chrysopidae

1,8-cineole, 315

citronellal, 327–328

click beetles. *See* Coleoptera: Elateridae

Clostera inclusa (poplar tentmaker), 266. *See also* Lepidoptera (moths): Notodontidae

Clostridium perfringens (bacterium), 25

coccinelline, 216, 217

cochineal bugs. *See* Hemiptera: Dactylopiidae

cochineal red, 132–136

cockroaches. *See* Dyctioptera

coiling, 38, 51, 59, 63, 234–235

Coleoptera (beetles): Anthicidae (flower), 225; Buprestidae (metallic wood borer), 196, 203–205; Cantharidae (soldier), 185–188; Carabidae (ground), 151–156, 157–162, 252, 253; Cerambycidae (long-horned), 196, 197–198, 231, 239–240; Chrysomelidae (leaf), 196, 244–249, 250–253, 255–260; Coccinellidae (ladybird), 119, 135–136, 206–210, 211–219, 287; Dermestidae (dermestid), 57; Dytiscidae (predaceous diving), 115–116, 153, 164, 168–172; Elateridae (click), 108, 199–202; Gyrinidae (whirligig), 153, 163–167, 168; Lampyridae (firefly), 170, 189–193, 196; Lycidae (net-winged), 194–198; Meloidae (blister), 196, 220–223, 224–227; Oedemeridae (fake blister), 227; Phengodidae (glowworm), 40–41; Pyrochroidae (fire-colored), 224–227; Scarabaeidae (scarabid), 205, 241–243; Silphidae (carrion), 116, 170, 173–177; Staphylinidae (rove), 178–184; Tenebrionidae (darkling), 208–209, 228–231, 232–235, 236–240

conifers, 318–319

Conus toxins, 18, 295

crayfish, 65

Creophilus maxillosus (hairy rove beetle), 178–184

p-cresol, 99–100

Crotalaria (rattlebox plants), 286–291

Cyanocitta cristata (blue jay), 98

cyanogenesis, 33–36, 43–47, 265, 273–276

Cybister (predacious diving beetles), 168. *See also* Coleoptera: Dytiscidae

cybisterone, 169–170

gum trees. *See* eucalypts
gyrinidal, 164

hairs, 55–58, 142–143; glandular, 213–
217, 306–308; urticating, 292–296
hairy rove beetle. *See Creophilus
maxillosus*
harvestmen. *See* Opiliones
head standing, 236–240
heat production, 157, 158, 160, 335
hedgehog, 311–312
Hemiptera (true bugs), 225;
Aleyrodidae (whiteflies), 57, 137–
140; Aphididae (aphids), 117–120,
121–124, 148–150; Belostomatidae
(giant water bugs), 113–116, 169;
Cercopidae (spittle bugs), 129–131;
Coreidae (leaf-footed bugs), 102–
107; Cydnidae (borrower bugs), 106;
Dactylopiidae (cochineal scale bugs),
132–136; Flatidae (plant hoppers),
125–128; Gerridae (water striders),
182; Pentatomidae (stink bugs),
105–106, 299; Reduviidae (assassin
bugs), 108–112, 248
Hemisphaerota cyanea (tortoise beetle),
244–249, 252–253. *See also*
Coleoptera: Chrysomelidae
Heterotheca psammophila (asteraceous
plant), 109, 110
n-hexyl acetate, 105
histamine, 324, 334
homoglomerin, 52
honey bees. *See* Hymenoptera: bees
hormones, 115–116, 170
hyaluronidase, 25, 334
hydrogen cyanide (HCN), 33–36, 43–
47, 273–276
hydrogen peroxide, 158–160
hydroxydanaidal, 286–291
(2-hydroxyethylamino)-alkadienoic
acid, 216, 217

(2-hydroxyethylamino)-alkenoic acid,
216, 217
9-(2-hydroxyethylamino)-decanoic
acid, 216, 217
8-(2-hydroxyethylamino)-nonanoic
acid, 216, 217
1-(2-hydroxyethyl)-2-(12-
aminotridecyl)-pyrrolidine, 212, 213
10-(2-hydroxyethylamino)-undecanoic
acid, 216, 217
2-hydroxy-3-methyl-1,4-
benzoquinone, 39, 40
15β-hydroxyprogesterone, 175–176
Hymenoptera (ants, bees, sawflies,
wasps), 225; ants, 5, 8, 13, 35, 46,
49–50, 53, 56–57, 63, 66, 75, 83,
84, 87, 91, 96, 109, 122, 126, 127,
135, 146, 150, 154, 155, 160, 174,
180, 186, 208, 209, 213, 216, 221,
233, 239, 245–246, 252, 253, 259,
262–263, 265, 269, 275, 279, 287,
307, 315, 321–330; bees, 23, 205,
331–337; sawflies, 262, 314–320;
wasps, 23, 63, 106, 118–119, 121,
122, 205, 280, 282, 335–336
Hyperaspis trifurcata (ladybird beetle),
135–136, 216. *See also* Coleoptera:
Coccinellidae

inchworms. *See Nemoria; Synchlora*
indole, 91–92
iridodial, 180–181
isoamyl acetate, 180–181, 335
isoamyl alcohol, 180–181
isobutyric acid, 298–299
isophorone, 99–100
isopropyl myristate, 146
Isoptera (termites), 82–88

jumping, 125–128, 129–131, 199–202

kempane, 83, 85

lacewings. *See* Neuroptera: Chrysopidae

ladybird beetles. *See* Coleoptera: Coccinellidae

Laetilia coccidivora (pyralid moth), 135

largemouth bass, 164–166

Latrodectus mactas (black widow spider), 24. *See also* Araneida

Lavandula luisieri (mint plant), 174

lavandulol, 174

Leiobunum nigripalpi (daddylonglegs), 11–14. *See also* Opiliones

Lepidoptera (butterflies): Lycaenidae (lycaenid), 123; Nymphalidae (brush-footed), 273, 309–313; Papilionidae (swallowtail), 297–303; Pieridae (whites and sulphur), 304–308

Lepidoptera (moths): Arctiidae (tiger), 286–291, 294; Dalceridae (dalcerid), 261–263; Geometridae (inchworm), 282–285; Lasiocampidae (lappet), 265; Limacodidae (slug caterpillar), 262, 294; Lymantriidae (tussock), 294; Megalopygidae (flannel), 294; Noctuidae (owlet), 264–268; Notodontidae (prominent), 266, 269–272; Oecophoridae (oecophorid), 265, 319; Pyralidae (pyralid), 135; Saturniidae (giant silkworm), 292–296; Sphingidae (sphinx), 208; Thyrididae (window-winged), 273–276; Yponomeutidae (ermine), 277–281; Zygaenidae (smoky), 273

Lethocerus americanus (giant water bug), 114. *See also* Hemiptera: Belostomatidae

Leucochrysa (lacewings): *floridana,* 147; *pavida,* 149–150. *See also* Neuroptera: Chrysopidae

Leucopis (flies), 135–136. *See also* Diptera

light emission, 35, 189–191

light sensing, in crayfish, 65–66

Limenitis archippus (viceroy butterfly), 309, 311. *See also* Lepidoptera (butterflies): Nymphalidae

limonene, 83, 85

Litoprosopus futilis (palmetto borer moth), 264–268

lizard, horned, 324–325

long-horned beetles. *See* Coleoptera: Cerambycidae

Lonomia achelous (saturniid moth), 294. *See also* Lepidoptera (moths): Saturniidae

Loxosceles (brown spiders): *laeta, reclusa, rufescens,* 24, 25. *See also* Araneida

lubber grasshoppers, 97–101

lucibufagin C, 190–192

lycidic acid, 196

Lycus fernandezi (net-winged beetle), 197–198. *See also* Coleoptera: Lycidae

Malacosoma americanum (eastern tent caterpillar), 265

mandelonitrile, 35–36, 45, 275

Manduca sexta (sphinx moth), 208

Mastigoproctus giganteus (vinegaroon or whipscorpion), 4–6

mayolene, 307

Megaloblatta blaberoides (cockroach), 66–67. *See also* Dyctioptera

melittin, 333

Metaleurodicus griseus (whitefly), 57, 137–140

methaqualone, 52, 53

3-methoxy-2,5-dimethyl-1,4-benzoquinone, 39–40

2-methoxy-3-isopropylpyrazine, 196

prevenom, 20

prickly pear cacti, 102, 103, 132, 133, 242

Prociphilus tessellatus (woolly alder aphid), 121–124, 148–150. *See also* Hemiptera: Aphididae

progesterone, 114–115

Prosapia bicincta (two-lined spittle bug), 129–131

prostaglandins, 25

pseudopelletierine, 213

Pseudoperga (sawflies), 315; *guerini*, 317–318. *See also* Hymenoptera: sawflies

pumiliotoxin-8, 324, 325

pyragines, 196–197

quinoline, 90, 92

reflex bleeding, 136, 195, 196–197, 212, 220–223

regurgitating, 8, 97, 99, 135–136, 265–267, 298, 316–320

repellency/deterrence: to birds, 52, 60, 95–96, 99, 154, 160, 174, 186, 216, 239, 305–306, 316; to fish, 115, 164–166, 169; to frogs, 60, 91, 160; to lizards, 60; to rodents, 5, 52, 60, 154, 186, 201, 239, 316; to toads, 44, 52–54

repellency/irritancy: to ants, 5, 8, 35, 46, 49–50, 75, 84, 91, 109, 127, 135, 146, 150, 154, 160, 174, 180, 186, 213, 216, 221, 239, 259, 265, 275, 287, 307; to other insects, 84, 127, 174, 186; to spiders, 8, 35, 66, 84, 86, 91, 127, 139, 142–143, 154, 176, 186, 191, 200–201, 213, 221, 286–287

Rhinocricus insulatus (millipede), 72. *See also* Diplopoda: Spirobolida

robustotoxin, 25

Romalea guttata (eastern lubber grasshopper), 97–101

rosemary bush. *See Ceratiola ericoides*

rove beetles. *See* Coleoptera: Staphylinidae

Sabal palmetto (cabbage palmetto), 125, 126, 253

Salicaceae (willow and poplar trees), 259

salicylaldehyde, 257

sawflies. *See* Hymenoptera: sawflies

scales, 57, 264

scarab beetles. *See* Coleoptera: Scarabaeidae

Schizura (unicorn caterpillar moths): *leptinoides, unicornis,* 269–272. *See also* Lepidoptera (moths): Notodontidae

Scolopendra heros (the giant Sonoran centipede), 29–32

Scorpiones (scorpions), 15–21

Scotch pine, 318

Scutigera coleoptrata (long-legged centipede), 31

secretory oozing, 8, 34, 37–42, 44, 48, 163–167, 168–172, 185–188, 242–243, 324

β-selinene, 299

selin-11-en-4α-ol, 299

Serenoa repens (saw palmetto), 125, 253, 265

Sicarius (spiders), 25. *See also* Araneida

Silpha (carrion beetles): *americana, novaboracensis,* 116, 170, 175, 176. *See also* Coleoptera: Silphidae

skatole, 142

skeletal toughness, 53–54, 233, 241

Solanum eleagnifolium (nightshade plant), 251

soldier beetles. *See* Coleoptera: Cantharidae